Lecture Notes in Civil Engineering 600

Lecture Notes in Civil Engineering (LNCE) publishes the latest developments in Civil Engineering—quickly, informally and in top quality. Though original research reported in proceedings and post-proceedings represents the core of LNCE, edited volumes of exceptionally high quality and interest may also be considered for publication. Volumes published in LNCE embrace all aspects and subfields of, as well as new challenges in, Civil Engineering. Topics in the series include:

- Construction and Structural Mechanics
- Building Materials
- Concrete, Steel and Timber Structures
- Geotechnical Engineering
- Earthquake Engineering
- Coastal Engineering
- Ocean and Offshore Engineering; Ships and Floating Structures
- Hydraulics, Hydrology and Water Resources Engineering
- Environmental Engineering and Sustainability
- Structural Health and Monitoring
- Surveying and Geographical Information Systems
- Indoor Environments
- Transportation and Traffic
- Risk Analysis
- Safety and Security

To submit a proposal or request further information, please contact the appropriate Springer Editor:

- Pierpaolo Riva at pierpaolo.riva@springer.com (Europe and Americas);
- Swati Meherishi at swati.meherishi@springer.com (Asia—except China, Australia, and New Zealand);
- Wayne Hu at wayne.hu@springer.com (China).

All books in the series now indexed by Scopus and EI Compendex database!

Edoardo Favari · Franca Cantoni · Marco Allena ·
Fabio Borghetti · Primiano Di Nauta
Editors

Sustaining the Future: Addressing Grand Challenges Through Megaprojects

MeRIT Workshop 2024

Springer

Editors
Edoardo Favari
Dipartimento di Ingegneria Gestionale
Politecnico di Milano
Milan, Italy

Franca Cantoni
Dipartimento di Scienze economiche e sociali
Università Cattolica del Sacro Cuore
Piacenza, Italy

Marco Allena
Dipartimento di Scienze giuridiche
Università Cattolica del Sacro Cuore
Piacenza, Italy

Fabio Borghetti
Dipartimento di Design
Politecnico di Milano
Milan, Italy

Primiano Di Nauta
Dipartimento di Economia
University of Foggia
Foggia, Italy

ISSN 2366-2557 ISSN 2366-2565 (electronic)
Lecture Notes in Civil Engineering
ISBN 978-3-031-92796-6 ISBN 978-3-031-92797-3 (eBook)
https://doi.org/10.1007/978-3-031-92797-3

This Springer imprint is published by the registered company Springer Nature Switzerland AG
The registered company address is: Gewerbestrasse 11, 6330 Cham, Switzerland

Preface

"Sustaining the Future: Addressing Grand Challenges through Megaprojects" highlights the pivotal role that megaprojects play in tackling some of the most pressing socio-economic and environmental challenges of our time. These projects—characterized by their vast scale, complexity, and transformative potential—serve as key drivers for innovation, infrastructure development, and sustainable progress. From climate resilience to urbanization, megaprojects are at the forefront of addressing the grand challenges that shape our global future.

However, the increasing complexity and interconnectivity of these challenges have also transformed how megaprojects are defined and conceptualized. No longer confined to traditional metrics such as budget thresholds or physical scale, contemporary megaprojects encompass a broader and more fluid array of goals. They aim not only to deliver tangible outcomes but also to advance sustainable development, foster social equity, and integrate cutting-edge technologies in highly uncertain environments. This evolution underscores the necessity for adaptive and systemic approaches in megaproject management, positioning these initiatives as dynamic, living systems. As such, the very definition of what constitutes a megaproject has become context-dependent, shaped by the interdependencies and scale of the grand challenges they seek to address. This fluidity reflects the transformative potential of megaprojects, reinforcing their critical role in sustaining the future.

This multifaceted nature of megaprojects demands a comprehensive exploration of their management frameworks, legal underpinnings, and practical applications, which this volume addresses through its three interconnected parts, each offering a unique lens through which the complexities of megaproject management are examined: management, legal, and case studies. Each part addresses different dimensions of the intricate world of megaprojects, ensuring a holistic approach to understanding and navigating these large-scale endeavors:

- the first section focuses on the management aspects of megaprojects. This section delves into the strategic, operational, and human resource dimensions, exploring leadership, soft skills, and organizational dynamics required to handle the immense scale and complexity of these projects. Emphasis is placed on the evolving role of project managers in a rapidly changing global environment, highlighting the importance of adaptability, communication, and stakeholder engagement;
- the second section covers the legal framework that governs megaprojects. Here, the focus is on regulatory compliance, risk management, and the contractual intricacies that define the legal landscape in which these projects operate. It explores how legal frameworks not only protect the interests of all stakeholders but also ensure that megaprojects adhere to international standards of sustainability, governance, and accountability;

- the third and final part, deliberately the most substantial of the volume, is dedicated to detailed case studies. This section is extensive because it provides in-depth, real-world examples of megaprojects from various industries and regions. The case studies bring theory to life, illustrating how the management principles and legal frameworks discussed earlier are applied in practice. By offering concrete examples, this section allows readers to gain practical insights into the challenges, successes, and lessons learned in managing complex megaprojects. The prominence of this part underscores the importance of empirical evidence in understanding how megaprojects are executed and governed in diverse contexts.

By structuring the volume in this way, we aim to provide readers with a comprehensive toolkit that combines theoretical knowledge, legal understanding, and practical insights, enabling them to better navigate the complexities inherent in megaproject management.

Section I. *Innovative Approaches and Dynamic Skills for Megaprojects Management*

The management and execution of megaprojects have increasingly become a focal point for addressing global grand challenges, such as sustainability, innovation, and large-scale infrastructure development. This collection of chapters delves into various critical dimensions of megaproject management, offering diverse perspectives that range from soft skills development to the role of artificial intelligence and governance in complex project environments.

The first chapter, *The Shapeshifter Model for Soft Skills for Project Managers* by Cantoni, Arcuri, and Di Nauta, explores the role of soft skills within megaproject management. The authors argue that unlike hard skills, soft skills are context-specific and dynamic, requiring project managers to continuously adapt to evolving cultural, social, and political landscapes. The proposed "Shapeshifter Model" offers a framework for assessing the maturity of these skills, emphasizing their importance in addressing the complexities of megaprojects and grand challenges.

In the second chapter, *Meta-Soft Skills Development in Megaproject Management* by Virtuani, Barabaschi, and Cantoni, the concept of meta-soft skills is introduced. These higher-order skills go beyond traditional soft skills, enabling project managers to integrate and coordinate complex interdependent sub-projects within a megaproject. The authors propose an educational model to cultivate these meta-skills, preparing leaders to handle the challenges of managing interconnected and large-scale initiatives.

The third chapter, *AI-enabled Multi-project Governance: A Proposed Framework* by Mariani et al., examines the potential of artificial intelligence in enhancing project governance in multi-project contexts. By integrating AI tools such as machine learning and predictive analytics, the chapter presents a governance framework that aims to improve decision-making and resource management in large-scale projects. This approach underscores the transformative potential of AI in handling the complexities of megaprojects.

In *Change of Expectations of the Socio-economic Context in the Evaluation of Megaprojects*, Bianchi explores the socio-economic changes affecting megaproject management, particularly the rise of BRICS and Arab countries as key players in global megaprojects. The chapter critiques the eclecticism seen in project management practices, arguing that unrealistic decision-making and over-reliance on interdisciplinary methods can hinder project success.

Finally, the chapter *Role of Megaprojects in Achieving Sustainability Transition* emphasizes the importance of governance, technological integration, and stakeholder engagement in driving the sustainability of megaprojects. By investigating the interplay between these factors, the study reveals how stakeholder engagement can mediate the effectiveness of governance and technological tools, ultimately improving project performance.

Together, these chapters provide a comprehensive examination of the multifaceted nature of megaproject management, exploring how leadership, soft skills, governance, technology, and socio-economic factors converge to shape the success of these large-scale initiatives.

Section II. *Legal and Regulatory Foundations for Megaprojects Execution*

The second section provides a thorough analysis of the legal frameworks that govern megaprojects. Megaprojects operate within intricate regulatory environments, where legal compliance, risk mitigation, and contractual governance are essential to their success. This section investigates how legal structures—ranging from public-private partnerships to environmental regulations—support or hinder the sustainability of megaprojects, emphasizing the need for robust legal governance to manage the long-term impacts of these projects on societies and ecosystems.

The sixth chapter, titled *The Reform of the Italian Tax System and Its Effect on the Implementation of Megaprojects* by Marco Allena, examines how recent reforms in Italy's tax system influence the execution of megaprojects. The chapter analyzes the implications of the changes to tax residency laws, the introduction of a global minimum tax, and reshoring incentives, focusing on how these reforms impact multinational corporations involved in large-scale projects. By offering a detailed review of the legislative changes, Allena highlights the critical role tax planning plays in ensuring the feasibility and financial sustainability of megaprojects.

In the seventh chapter, *The New "Right of Questioning" and Its Effects for the Megaprojects Sector* by Andrea Purpura, the focus shifts to the Italian "right of questioning" within the tax system. Purpura explains how this reform promotes early engagement between large taxpayers, such as those involved in megaprojects, and the Italian Revenue Agency. The chapter underscores how this structured dialogue fosters legal certainty, reduces the likelihood of disputes, and provides a more efficient pathway for resolving complex tax issues, which is especially beneficial in the high-stakes megaproject sector.

In the eight chapter, *Negotiating Megaprojects and Good Faith* by Francesco Zecchin, the focus is on the legal principle of good faith in the negotiation processes of megaproject contracts. Zecchin analyzes the implications of precontractual liability and examines what happens when good faith is breached in both public and private sector negotiations. This chapter emphasizes the importance of transparency and fairness in contract negotiations to prevent disputes and ensure the smooth progression of megaprojects.

The ninth chapter, *Mega-Projects and Permanent Establishment: The Relevance of the Negative List Carve-Outs and the Fragmentation of Activities* by Paolo Arginelli and Francesco Reboli, explores the evolving concept of permanent establishment (PE) in the context of international tax law. The authors discuss the relevance of the negative list and how the fragmentation of activities under the BEPS project impacts the taxation of cross-border megaprojects. This chapter is essential for understanding how tax jurisdictions handle the complexities of international megaprojects, offering strategies for minimizing legal and financial risks

Collectively, these chapters offer a comprehensive analysis of the critical challenges inherent in the management of megaprojects. Through rigorous legal examination and evidence-based insights, this volume provides stakeholders with the essential knowledge and frameworks needed to effectively navigate the complex regulatory landscapes that shape the success and sustainability of megaprojects.

Section III. *Lessons from Leading Cases*

The third and most substantial section is dedicated to detailed case studies that bring the theoretical and legal discussions to life. Through these case studies, the book illustrates how megaprojects around the world are responding to grand challenges in areas such as renewable energy, transportation infrastructure, and urban development. This section's prominence reflects the need for empirical evidence and practical insights, showing how megaprojects can be both innovative solutions and cautionary tales in addressing global challenges. These examples reveal the lessons learned, best practices, and pitfalls encountered in managing such ambitious undertakings, providing readers with actionable insights into the complex world of megaproject execution.

The tenth chapter, Is a Hyperloop System a Megaproject? An Italian Case Study between the Cities of Milan and Rome by Fabio Borghetti, Federico Ursino, and Michela Vischi, explores the feasibility of implementing a Hyperloop transportation system in Italy. The chapter examines the financial, technical, and socio-economic challenges of this ambitious megaproject, considering its potential long-term economic benefits despite high initial costs. Through detailed financial assessments and revenue scenarios, the study highlights the transformative potential of the Hyperloop in revolutionizing mobility between Milan and Rome.

The eleventh chapter, Sustainability Disclosure and Legacy of Sports Mega-Events: The Case of Paris 2024 Olympic Games by Andrea Caccialanza, Ernesto De Nito, and Paolo Canonico, investigates the sustainability and legacy management of the Paris 2024 Olympic Games. Through a qualitative analysis of interim sustainability and legacy reports, this chapter examines the challenges of ensuring transparency and accountability in mega-events, with a focus on stakeholder engagement and the long-term social and environmental impacts of such a large-scale project. The chapter adopts a framework that assesses stakeholder engagement across various phases of the project and offers critical insights into the effectiveness of sustainability practices in addressing the complex legacy of the Games

The twelfth chapter, titled *Infrastructure Refurbishment – The Tunnel Renewal Strategy: From Knowledge to Renovation of the Asset* by Daniele Di Fiore, Carlo Alessio, Lapo Baccolini, Fabrizio Colla, and Marilisa Conte, explores the innovative approach undertaken by Autostrade per l'Italia for the refurbishment and renewal of existing tunnels within the Italian motorway network. The Tunnel Renewal Strategy (TRS) is a comprehensive program designed to extend the lifespan of tunnels by at least 50 years through structural reinforcement and modernization efforts. This chapter presents detailed case studies of various tunnel renewal projects, highlighting the implementation of high-performance materials, advanced construction techniques, and sustainability practices. By integrating state-of-the-art engineering solutions and environmental considerations, the TRS aims to enhance both the safety and resilience of Italy's aging tunnel infrastructure.

In the thirteenth chapter, *Innovative Project Management for Sustainable Infrastructure: Coastal Highway Upgrade in Belize* by Gori et al., the focus shifts to sustainable infrastructure development. The chapter showcases the use of advanced engineering solutions and environmental management strategies in upgrading a flood-prone highway. The project integrates geosynthetic materials, bioengineering techniques, and solar-powered systems, aligning with Belize's renewable energy goals and enhancing the sustainability of its infrastructure.

The fourteenth chapter, *Transforming Megaprojects: Insights from High-Speed Rail Development* by Jithin Arakkatt Shaji and Ritwik Aravindakshan, investigates the evolution of high-speed rail megaprojects. The authors analyze how these projects, which often span decades, adapt to changing technological and regulatory environments. By focusing on projects like the High-Speed 2 (HS2) in the UK and other global initiatives, the chapter provides a critical look at the factors driving scope changes and budget overruns in long-term megaprojects.

The fifteenth chapter, *Mega Science and Re-Nationalization in the Field of Semiconductors: An Evidence-Based Exploration of Decoupling Triggered by the China Initiative* by Giulio Marini and Sara Lombardi, examines the geopolitical implications of megaprojects in the semiconductor industry. Focusing on the decoupling between China and the US in scientific collaborations, the chapter analyzes how national policies like the China Initiative are reshaping global research efforts and impacting the production of cutting-edge technologies.

The sixteenth chapter, titled *Using an Allocation Algorithm to Analyse Access to Cardiology Services in Italy* by Domenico Marino, Giuseppe Quattrone, and Francesco Timpano, presents a study focused on improving healthcare resource allocation, particularly cardiology services, across Italy. The chapter introduces an allocation algorithm implemented in R to simulate patient flow from municipalities with populations over the age of 65 to hospital facilities specializing in cardiology. By integrating demand distribution, transport distances, and hospital availability, the model aims to optimize access to healthcare services, reducing inequalities between urban and rural areas. The analysis provides valuable insights into improving patient flow management and offers recommendations for the strategic planning of healthcare resources based on geographic and demographic factors.

By integrating managerial, legal, and practical perspectives, Sustaining the Future: Addressing Grand Challenges through Megaprojects offers a comprehensive view of how these large-scale initiatives can shape a sustainable and resilient future. It equips readers with the tools and knowledge necessary to lead, manage, and govern megaprojects in ways that contribute to the long-term well-being of communities and the planet.

The Editors

Contents

Lessons from Leading Cases

Innovative Approaches and Dynamic Skills for Megaprojects Management

The Shapeshifter Model for Soft Skills for Project Managers: A Dynamic and Context-Sensitive Approach to Assessing Adherence and Maturity Levels in Addressing "*Grand Challenges*"

Franca Cantoni[1](✉), Marco Arcuri[2], and Primiano Di Nauta[2]

[1] Università Cattolica del Sacro Cuore, Piacenza, Italy
franca.cantoni@unicatt.it
[2] Università degli Studi di Foggia, Foggia, Italy

Abstract. Unlike hard skills, which are quantifiable, delineating the optimal soft skills profile for a project manager (PM) presents a multifaceted challenge due to their multifarious nature. Cultural, social, and political-institutional contexts, along with the dynamic nature of projects, play a critical role in influencing situations that activate and shape the manifestation of soft skills. These factors interact with individual differences and interpersonal interactions, thereby affecting how soft skills are expressed and perceived in various project environments. Furthermore, the diversity among stakeholders, including their differing expectations, communication styles, needs, and behaviors, can significantly impact how soft skills manifest. In response to the growing complexity of grand challenges—global, multifaceted problems that require interdisciplinary collaboration—an off-beat approach based on multidimensional skills integration (MSI), contextual embeddedness (CE), project-specific adaptability (PSA), situational responsiveness (SR), and stakeholders' diversity (SD) is developed. This framework highlights how multilayer and dynamic feedback loops (MDFL) can generate the activation of various soft skills with distinct nuances and shape. By incorporating all these variables, the "shapeshifter model" offers a dynamic and accurate assessment of soft skills, tailored to the different situations that manifest in each project. Moreover, to address the complex nature of soft skills development and its contextual manifestations, this study introduces a novel maturity model that provides a more precise and context-sensitive understanding of soft skills proficiency, catering to the evolving demands of modern project environments, especially in the context of grand challenges.

Keywords: Soft skills · assessment · context · interpersonal interactions · individual differences · maturity model

E. Favari et al. (Eds.): MeRIT 2024, LNCE 600, pp. 3–19, 2025.
https://doi.org/10.1007/978-3-031-92797-3_1

1 Introduction

Traditionally, PM was largely synonymous with technical prowess and proficiency in the intricacies of project methodologies. The emphasis was placed squarely on the mastery of tools, processes, and techniques, often overlooking the human element [1]. However, the recognition that projects are fundamentally social endeavors, shaped by the collaboration and interaction of diverse individuals [2, 15] in ever-changing organizational landscapes [19, 20] has given way to a more holistic understanding that integrates technical competence with a rich tapestry of soft skills [11, 27]. Consequently, there has been a growing awareness that delineating the ideal profile for a project manager who can effectively contribute to project success or seamlessly integrate within a team dynamic involves a primary challenge that extends beyond merely assessing technical proficiency. Rather, it requires a comprehensive understanding of their repertoire of soft skills [17].

Unfortunately, unlike hard skills, which are characterized by their quantifiability and the availability of standardized assessment methods, crafting soft skills—both in general and specifically within the realm of PM — poses a complex challenge [5, 9, 14]. One primary impediment lies in their intangible nature [22]. While hard skills are delineated by specific tasks, procedures, or knowledge domains, soft skills are often manifested through nuanced behaviors and socio-emotional competencies [7, 18]. Secondly, by definition, soft skills are context-specific, meaning that they can manifest differently depending on the context in which they are activated [12, 13, 16, 21, 19]. General context in PM plays a critical role, influencing every stage of a project's lifecycle [3, 24, 29]. It is multifaceted, encompassing various dimensions such as political/institutional, social, cultural and project-specific elements. Political factors, including government policies and political stability, can influence project feasibility, funding, and risk management. Political changes may lead to shifts in priorities or regulatory frameworks, requiring project managers to remain adaptable and informed. Engagement with political entities and advocacy groups can also affect project approval and support, particularly in the public sector.

On the other hand, the socio-cultural context encompasses social norms, values, and cultural differences that can affect team dynamics, communication, and collaboration. Managing these differences is vital for fostering effective teamwork and ensuring project success. Additionally, considering the impact of the project on the community and environment is essential for sustainable project management, aligning with broader societal values and corporate social responsibility.

Additionally, the project-specific context should not be overlooked as it involves understanding the unique goals, scope, and constraints of each project.

Recognizing and aligning with these contextual factors, both general and project-specific, is essential for effective project planning, execution, and achieving successful outcomes. A comprehensive understanding of the context enables project managers to anticipate challenges, capitalize on opportunities, and make informed decisions that align with overarching strategic objectives.

Furthermore, it is crucial to recognize that contextual specificities—cultural, institutional, and social—along with the unique characteristics of each project, are dynamic and evolve over time. As the project advances through different phases, the context can shift—new stakeholders may become involved, external conditions may change, or internal team dynamics may evolve. Project managers must remain vigilant and adaptable, continuously reassessing and applying their soft skills to meet the shifting demands of the project environment.

This dynamic interplay results in continuous mutual influence between context and project-specific factors, creating a fluid environment. Consequently, situations arise that are both shaped by and contributory to the evolving context.

In this fluid environment, the interaction between context and project specifics leads to the continuous alignment of soft skills. This reciprocal influence results in a multilayer and dynamic landscape where project managers must constantly adjust their approach, leveraging their soft skills to navigate the evolving conditions effectively.

It is precisely the context, both general and specific, that generates a multitude of situations, activating the PM's soft skills in varied ways, which we might describe as "situational." This comprehensive awareness and adaptability are essential for navigating the complexities of modern PM, as the same soft skill may exhibit different characteristics, behaviors, or impacts depending on the context that shapes and defines each situation. For example, communication skills might look different in an international team setting compared to a client meeting, or problem-solving skills might vary between a high-pressure emergency and a routine task. Understanding this context-specific nature of soft skills is crucial for accurately assessing them within different settings.

Furthermore, individual differences [6] —such as personal growth and development—and interpersonal interactions [18]—such as team dynamics and the leadership style of project leaders—can give rise to micro-level dynamics that differentially activate and craft soft skills. The interplay between individual differences and interpersonal interactions contributes to a perpetually evolving the soft skills set. As team members engage and develop, the requisite soft skills for navigating these interactions also transform. This continuous evolution necessitates a PM's ongoing adaptation and flexibility in applying their soft skills.

The dynamic nature of these micro-level interactions means that soft skills are not static but are activated differently depending on the situation. For example, during the initial stages of a project, communication and negotiation skills may be paramount as the team aligns on goals and expectations. As the project progresses, leadership and conflict resolution skills might become more critical to maintain momentum and address emerging challenges.

In summary, the complex interplay between individual differences, interpersonal interactions, and evolving contextual and project-specific factors creates a dynamic environment where soft skills are continually activated and redefined. Understanding and navigating these micro-level dynamics is essential for project managers to achieve successful project outcomes.

This study addresses the research gap by challenging the "principle of homogeneity" and rejecting static approaches, recognizing the different nuances and shapes soft skills can assume by highlighting the critical influence of context and stakeholders' diversity, which traditional approaches have overlooked. To bridge this gap, we propose a novel model offering a more accurate and dynamic crafting tailored to the unique environments of each project.

To address these challenges, our study delineates two primary objectives. Primarily, we aim to circumvent conventional approaches that rely on static consideration and predefined criteria for analyzing soft skills. Traditional methodologies often fail to capture the dynamic, multifaceted, and context-dependent nature of soft skills. Secondly - recognizing the crucial role played by the context and stakeholders' diversity - we propose an architectural model designed to make the crafting of soft skills context and stakeholders'sensitive. This model aims to enhance the precision and relevance of soft skill assessment by accounting for the specific situational factors that influence their manifestation and impact.

The originality of this contribution is evident in its novel approach to soft skill consideration, which moves beyond traditional, one-size-fits-all methods.

Moreover, it contributes to the literature on "grand challenges" by offering a novel approach to addressing the complexities of project management in dynamic and context-sensitive environments. By focusing on the development and assessment of soft skills in project managers through the "Shapeshifter Model," the paper highlights the importance of adaptability, situational responsiveness, and stakeholder diversity—key elements in tackling grand challenges. These challenges, which often span across cultural, political, and institutional contexts, require nuanced and multifaceted leadership skills. The paper's integration of dynamic feedback loops and context-specific soft skills assessment frameworks provides a fresh perspective on how project managers can navigate and contribute to solving complex, global-scale problems, enhancing both the theoretical understanding and practical implementation of soft skills [4,8, 23, 26] in "grand challenges".

The paper is conceptual in nature and, therefore, no new empirical evidence is reported. Particular attention is paid to suggesting avenues for further research based on the consideration of soft skills based on a dynamic and context-specific environment.

The structure of this paper is delineated as follows. Following this introduction, Sect. 2 provides a literature review on traditional approaches for considering soft skills in project management and the role of the context. Additionally, it delineates the research proposition guiding this study. Subsequently, Sect. 3 elucidates the conceptual basis and rationale for model development. Following this, Sect. 4 expounds upon the conceptual model, which is subsequently elaborated in Sect. 5 with the proposal of a maturity model for assessing soft skills in PMs and discussed in Sect. 6 offering a new perspective on soft skills consideration. Conclusions in Sect. 7 and their implications in Sect. 8 ensue, culminating in an exploration of limitations and avenues for further research.

2 Literature Review

Conventional approaches to evaluating soft skills for PMs exhibit significant limitations, particularly rational methodologies based on fundamental assumptions that fail to capture their complexity and nuances.

Most of the conventional studies consider soft skills are single-faceted and defined a priori [5]. Moreover, conventional assessments often rely on the '**assumption of homogeneity**,' presuming that soft skills are uniform and can be applied consistently across different individuals and situations [22]. This oversimplified approach reduces soft skills to single-dimensional constructs, failing to capture their complexity and context-dependent nature. As a result, such assessments often do not accurately reflect the nuanced and multifaceted reality of soft skills in diverse project environments. For example, assessing teamwork solely based on observable interactions overlooks deeper components such as conflict resolution, delegation, and fostering a collaborative environment. This oversimplification results in assessments that do not fully reflect a PM's proficiency in real-world scenarios.

Other studies (Strohmeier, 1992; [15, 25]) often fail to adequately consider or marginally consider **'interpersonal differences'** among PMs. These individual differences can profoundly impact the expression and perception of soft skills, leading to significant variations in how these skills are demonstrated and interpreted in various contexts. Factors such as personality traits, cultural background, and previous experiences play a crucial role in shaping an individual's soft skills. Ignoring these individual differences leads to a one-size-fits-all evaluation that may not accurately reflect each PM's unique abilities.

Traditional assessments often overlook the aspect of **'personal growth and development'** over time [10]. Soft skills are not static; they evolve with experience, feedback, and continuous learning. An assessment that provides only a snapshot at a single point in time fails to capture this dynamic progression, potentially undervaluing the developmental trajectory of a PM's soft skills.

Additionally, the **'style and approach of project managers'** can greatly affect which soft skills are most needed and valued within a team. For instance, a transformational PM may emphasize inspiration and motivation, while a transactional one may focus more on structure and efficiency. Assessments that do not consider these variations in leadership style and approach may fail to identify the specific soft skills that are most relevant and impactful in different leadership contexts.

Therefore, an effective crafting system should incorporate mechanisms to account for this ongoing development and the interplay between personal growth and development, the style and approach of project managers and contextual changes. Indeed, most approaches used to assess soft skills do not fully or adequately consider the role of context, which can be defined according to a multitude of facets, as highlighted by several authoritative scholars.

Despite the critical importance of these factors, current literature predominantly presents fragmented and isolated approaches that fail to integrate the broader and specific contexts, stakeholder diversity, interpersonal differences, and personal growth into a cohesive system. This gap highlights the need for more comprehensive and systematic method in crafting soft skills of project managers.

From these foundations, we derive our research propositions, which are as follows:

Proposition 1: Traditional static evaluation approaches are inadequate in dynamic project contexts, as they fail to capture the nuanced and context-dependent nature of soft skills.
Proposition 2: Contextual factors critically influence the project-specific environment, and, reciprocally, the project-specific environment shapes these contextual factors.
Proposition 3: The mutual interaction between the general context and the specific project environment both affects and is affected by situational dynamics.

Proposition 4: Micro-level dynamics, including individual differences and interpersonal interactions, can significantly influence the manifestation and shape of soft skills.
Proposition 5: Stakeholders' diversity can significantly impact the manifestation and shape (crafting) of soft skills.

3 Conceptual Basis and Rationale for Model Development

The methodology used to explain the Soft Skills Shapeshifter Model (3SM) is based on several well-defined "core concepts", including:

Multidimensional Skill Integration (MSI): PMs must cultivate a diverse set of soft skills across various domains to thrive in complex and rapidly changing environments. This concept supports the need for PM to develop a comprehensive soft skill set. The ability to integrate the soft skills effectively ensures that employees are well-rounded and capable of addressing a broad spectrum of challenges;
Contextual Embeddedness (CE): projects operate within and are profoundly influenced by multiple contextual layers, including political, institutional, social, and cultural environments. Understanding and adapting to these external forces is crucial for strategic planning and decision-making. The model's emphasis on political, institutional, social, and cultural factors aligns with this concept, highlighting the importance of contextual awareness in PM operations;
Project-Specific Adaptability (PSA): each project presents unique demands and challenges that require tailored strategies for successful execution. This concept underscores the need for flexibility in processes and resource allocation, allowing organizations to customize their approaches based on the specific requirements and constraints of each project. This adaptability is crucial for achieving project objectives and maintaining overall organizational effectiveness.
Situational Responsiveness (SR): soft skills must be applied differently depending on the immediate circumstances and challenges faced in a given situation. This concept emphasizes the need for project managers to be responsive and adaptable, adjusting their soft skills to effectively address situational demands. It highlights the dynamic nature of soft skills and their context-dependent application.
Stakeholders' Diversity (SD): the diversity among stakeholders, including their differing expectations, communication styles, needs, and behaviors, can significantly impact how soft skills are manifested. This concept focuses on the necessity for project managers to recognize and navigate the varied perspectives and influences of diverse stakeholders to ensure successful collaboration and project outcomes.
Dynamic Feedback Loops (DFL): Continuous feedback mechanisms are essential for enabling real-time adjustments and fostering a culture of ongoing learning within the project. By incorporating dynamic feedback loops, organizations can ensure that their skills and contextual understanding are constantly updated and refined. This continuous improvement process is vital for maintaining relevance and competitiveness in a rapidly evolving landscape.

The integration of the core concepts within the Soft Skills Shapeshifter Model (3SM) is designed to create a cohesive and dynamic model. Each concept interlinks with others, facilitated by multiple layers of Dynamic Feedback Loops (DFLs), which enable continuous learning, adaptation, and improvement. The integration is detailed in multiple layers as follows:

1st layer: CE emphasizes the importance of context awareness and adaptability. The interplay between the different factors composing the CE ensures that PMs not only possess the necessary skills but also understand how to apply them within the specific contexts they operate. This contextual adaptation is crucial for strategic decision-making and planning;
2nd layer: PSA adds a layer of flexibility to the model. It highlights the need for PMs to tailor their approaches to the unique demands and constraints of each project. The dynamic nature of projects requires continuous recalibration of strategies and resource allocation. By integrating PSA, 3SM ensures that PMs can pivot and adapt their approaches to meet the specific needs of each project, thereby enhancing the likelihood of successful outcomes;
3rd layer: SR emphasizes the need for real-time adaptability in the application of soft skills. This layer focuses on the immediate and changing demands of project environments. By fostering situational responsiveness, PMs can adjust their behavior and strategies to effectively address emerging challenges. SR works synergistically with PSA, ensuring that PMs remain agile and responsive, thereby optimizing their performance in dynamic situations;
4th layer: MSI serves as the foundational layer of 3SM. By emphasizing the cultivation of a diverse set of soft skills, MSI ensures that PMs are equipped with a broad and versatile skillset. This multidimensional approach allows PMs to handle a variety of challenges effectively. The integration of MSI into 3SM provides the basis for developing well-rounded individuals capable of seamless skill application across different project phases and contexts;
5th layer: SD addresses the complexity introduced by diverse stakeholder groups. By recognizing and navigating the varied expectations, communication styles, and needs of stakeholders, PMs can facilitate better collaboration and project outcomes. The integration of SD within 3SM ensures that PMs can manage stakeholder relationships effectively, leveraging diversity to enhance project performance. This focus on stakeholder diversity also complements the other core concepts by providing a comprehensive understanding of the human element in project management.

DFLs act as the integrating mechanism that weaves together all the core concepts. By incorporating continuous feedback mechanisms, 3SM ensures that PMs and their teams can learn and adapt in real time. This ongoing process of feedback and adjustment will allow for the refinement of soft skills and contextual understanding. DFLs will enable the model to remain dynamic and responsive to changes, thereby maintaining relevance and fostering continuous improvement.

The synthesis of these core concepts within 3SM will create a robust and flexible framework for project management. Each concept is interconnected, allowing for a holistic approach to developing and applying soft skills. The dynamic interplay between

MSI, CE, PSA, SR, SD, and DFLs will ensure that PMs will be well-equipped to handle the complexities of modern project environments.

4 The Conceptual Model

As shown in Fig. 1 the Soft Skills Shapeshifter Model (3SM) integrates several core concepts into a cohesive theoretical framework that emphasizes the dynamic interplay between soft skills, situational demands, project-specific elements, and broader contextual factors. The methodology employed reveals significant insights into how these elements interact to shape the application and efficacy of soft skills in project management.

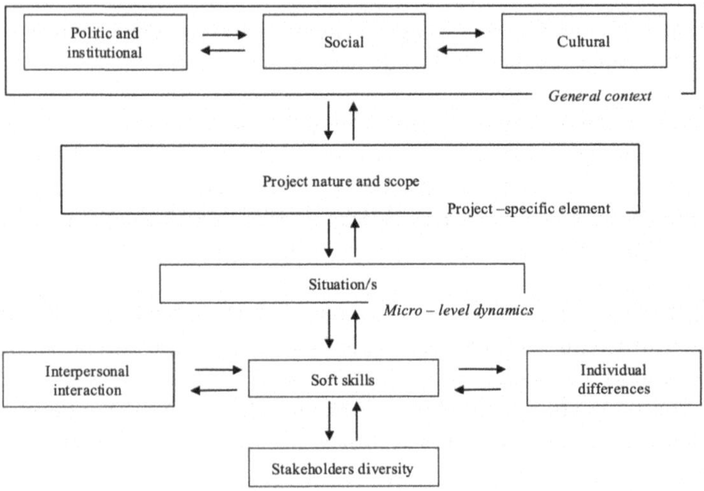

Fig. 1. The Soft Skills Shapeshifter Model (3SM)

Central to the 3SM is the concept of CE, which illustrates how political, institutional, social, and cultural factors profoundly influence project-specific elements and the deployment of soft skills. The broader context shapes the environment in which projects are executed, necessitating project managers to navigate complex frameworks and adapt their skills accordingly. For instance, an organization in a highly regulated industry must prioritize compliance and ethical decision-making, while cultural norms influence team interactions and leadership styles, affecting how soft skills are applied.

PSA further underscores the need for tailored strategies to address unique project demands. Each project's specific goals, timelines, resources, and stakeholder expectations create a distinct situational context that shapes the application of soft skills. Projects with tight deadlines, for example, demand exceptional teamwork and time management skills, whereas those requiring extensive stakeholder engagement highlight the importance of negotiation and communication skills. This adaptability is crucial for achieving project objectives and maintaining organizational effectiveness.

The efficacy of soft skills is also significantly influenced by SR. Situational factors such as the immediate environment, team dynamics, and specific challenges necessitate that project managers dynamically adjust their soft skills. High-pressure situations may require enhanced problem-solving abilities and adaptability, demonstrating that the situational context can greatly affect the effectiveness of soft skills. Figure 2 shows the impact of different situations on soft skills.

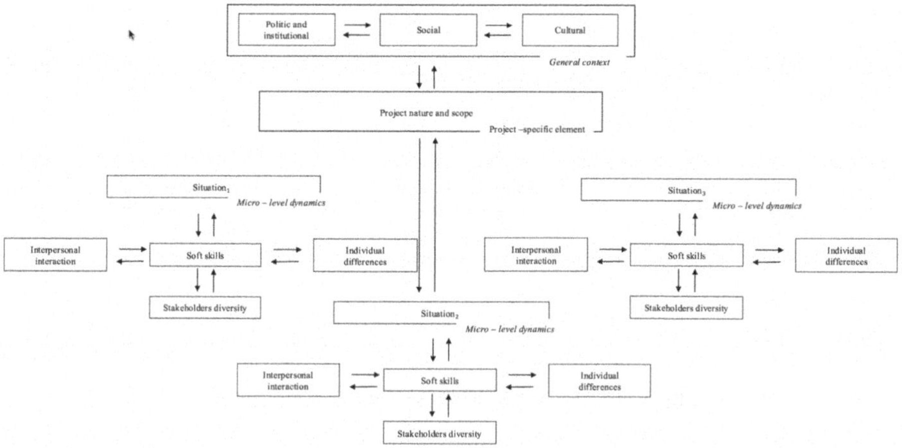

Fig. 2. An example of the Soft Skills Shapeshifter Model (3SM) applied to three different situations

Moreover, the diversity among stakeholders—encompassing differing expectations, communication styles, needs, and behaviors—plays a critical role in how soft skills are manifested and applied. Recognizing and navigating these varied perspectives is essential for successful collaboration and project outcomes. Effective management of stakeholder diversity enhances team cohesion, communication, and overall project success.

Continuous feedback mechanisms, encapsulated in the concept of DFL, are vital for real-time adjustments and fostering ongoing learning within projects. Incorporating dynamic feedback loops ensures that skills and contextual understanding are constantly updated and refined. This continuous improvement process is essential for maintaining relevance and competitiveness in a rapidly evolving landscape, enabling project managers to adapt their strategies proactively.

Finally, the integration of multidimensional soft skills, as outlined in MSI, ensures that employees are well-rounded and capable of addressing a broad spectrum of challenges. Effective communication within a team, for instance, can vary significantly depending on the project's specific requirements and the overall cultural context of the organization. The 3SM ensures that employees are equipped to handle diverse tasks and challenges, promoting alignment, flexibility, and adaptability. Figure 3 shows an overview on how the core concepts interact.

In conclusion, the 3SM provides a comprehensive framework for understanding the dynamic interplay between soft skills, situational factors, project-specific elements, broader contextual influences and stakeholders' diversity. This model underscores the

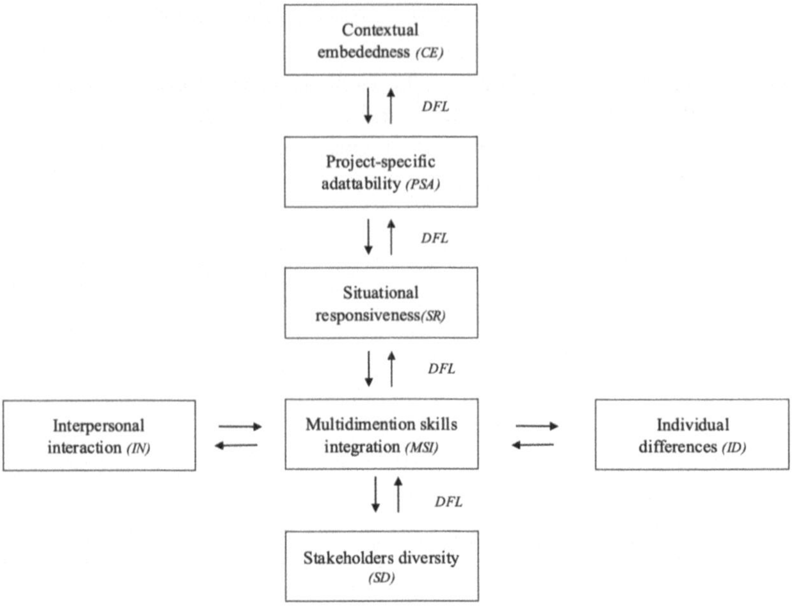

Fig. 3. CE, PSA, SR, MSI, SD and DFL in action

necessity for PM to be contextually aware, adaptable, and responsive to the unique demands of each project. By integrating these core concepts, the 3SM promotes a nuanced approach to soft skill development, ensuring that PMS are well-prepared to navigate the complexities of modern project environments effectively.

5 Proposal of a Maturity Model for Assessing Soft Skills in Project Managers

To conduct a thorough assessment of the various dimensions within the soft skills maturity model for project managers, a range of rigorous tools and methodologies should be employed, allowing for systematic data collection, precise analysis, and the generation of actionable feedback across multiple competencies.

For Contextual Embeddedness (CE), structured self-assessment questionnaires combined with third-party feedback offer a robust mechanism to evaluate a project manager's ability to adapt to political, institutional, social, and cultural contexts. The use of 360-degree feedback tools provides a comprehensive perspective on their adaptability across different settings. Scenario simulations, designed to mirror complex real-world situations, are instrumental in assessing how project managers react and adapt. Additionally, analyzing case studies of past projects enables evaluators to observe how managers handled contextual changes and adapted to new environments.

In assessing Project-Specific Adaptability (PSA), a detailed review of past project outcomes is essential to determine how project managers tailored their strategies and resources to meet the unique needs of each project. Interviews and focus groups with team members and stakeholders yield direct feedback on how well the manager customized strategies and managed resources. Furthermore, an analysis of planning documents can reveal how effectively project plans were adapted to address specific challenges and requirements.

For Situational Responsiveness (SR), simulations and role-playing exercises are highly effective in testing a project manager's ability to respond in real-time to unexpected events and emerging challenges. Keeping an incident and response log provides a detailed record of how managers handled crises and high-pressure situations, offering insights into their decision-making under stress. Additionally, specific assessments designed to measure performance in high-stress scenarios can provide valuable data on their ability to manage pressure and respond to sudden challenges.

Evaluating Stakeholder Diversity (SD) requires the administration of stakeholder feedback questionnaires to capture diverse perspectives on the project manager's ability to navigate varying expectations and communication styles. A comprehensive analysis of stakeholder relationship management strategies can reveal how effectively the project manager builds and maintains relationships with diverse groups. Additionally, brainstorming sessions and workshops with stakeholders offer a platform to observe how the project manager facilitates collaboration and resolves conflicts among stakeholders with differing interests.

The Interpersonal Interaction (II) dimension can be assessed through direct observation of the project manager's daily interactions, especially during meetings and discussions, to gauge their communication and conflict management skills. 360-degree feedback collected from colleagues, supervisors, and subordinates provides a well-rounded view of the project manager's interpersonal abilities. Role-playing and simulations, where managers must handle conflict, negotiation, and collaboration in controlled settings, further evaluate their proficiency in these areas.

For Individual Differences (ID), psychometric tools such as personality tests are utilized to assess traits like openness to change, resilience, and stress management. Cognitive assessments are also crucial in measuring key abilities such as critical thinking and problem-solving skills. Encouraging self-reflection and self-assessment allows project managers to analyze how their personal preferences and styles influence their behavior and decision-making processes.

In assessing Multidimensional Skill Integration (MSI), self-assessment questionnaires are used to evaluate how project managers apply various soft skills—such as communication, leadership, and problem-solving—in different project situations. Peer evaluations further contribute to understanding how these skills are integrated and applied in practice. Case studies of past projects, along with simulations, provide a deeper examination of how project managers combine their soft skills to achieve project objectives. Continuous monitoring of daily interactions, combined with real-time feedback during project phases, offers valuable insights into how managers integrate these skills into their routine activities.

Finally, for Dynamic Feedback and Continuous Improvement (DFCI), 360-degree feedback systems facilitate the collection of real-time input from colleagues, supervisors, and team members, enabling project managers to make continuous improvements. Performance monitoring tools, including software designed to track soft skill performance over time, allow for the identification of trends and areas that require further development. Encouraging regular reflection and self-assessment sessions supports project managers in identifying opportunities for personal growth and setting developmental goals.

By employing these rigorous assessment tools and methodologies across the eight dimensions, a comprehensive understanding of a project manager's soft skills maturity can be achieved, guiding both individual development and broader organizational improvement efforts.

The model showed in Fig. 4 measures adherence to each dimension with a percentage-based assessment. Additionally, individual assessments are weighted and normalized to contribute to the overall maturity level, measured on a scale from 1 to 5, termed: Initial, Basic, Medium, Advanced, Optimal.

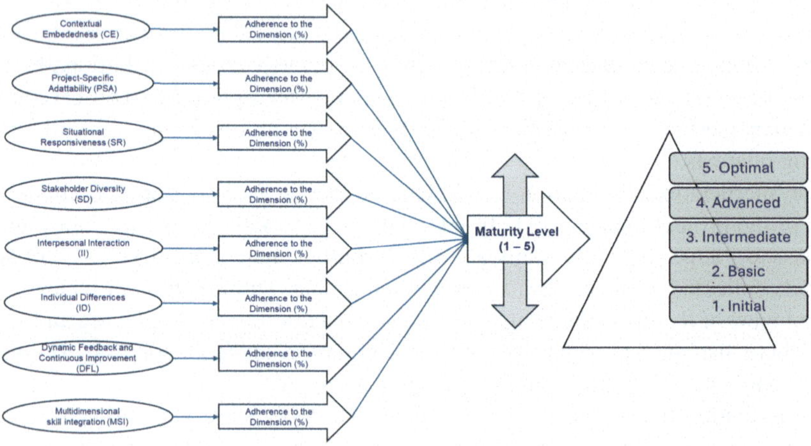

Fig. 4. Assessment of Adherence to Dimensions and Maturity Level

This multidimensional approach allows for a more in-depth and nuanced evaluation of soft skills, overcoming the limitations of traditional models that often offer a static and poorly contextualized view. The proposed methodology provides a more accurate and dynamic representation of competencies, reflecting the real operational and strategic needs of projects.

6 Discussion

Our research underscores the profound impact of contextual factors—including political, institutional, social, and cultural elements—on the application of soft skills. These factors necessitate a high degree of adaptability from project managers. This finding aligns with previous research that emphasizes the necessity for project managers to be contextually aware and flexible in their approach.

Furthermore, the need for tailored strategies in response to unique project demands is another significant insight. Project-specific elements such as goals, timelines, resources, and stakeholder expectations create distinct situational contexts that shape the application of soft skills. Projects with tight deadlines require different soft skills compared to those involving extensive stakeholder engagement. This adaptability is essential for achieving project objectives and maintaining organizational effectiveness, reinforcing the idea that one-size-fits-all approaches to soft skills assessment are inadequate.

The efficacy of soft skills is heavily influenced by situational factors. High-pressure environments, team dynamics, and specific project challenges necessitate that project managers dynamically adjust their soft skills. This responsiveness is crucial as it allows project managers to effectively address the immediate needs of the project. Our findings highlight the importance of situational context in determining the effectiveness of soft skills, suggesting that static assessments fail to capture the fluid nature of project environments.

Additionally, the diversity among stakeholders—encompassing differing expectations, communication styles, needs, and behaviors—plays a critical role in how soft skills are manifested and applied. Recognizing and navigating these varied perspectives is essential for successful collaboration and project outcomes. This insight underscores the importance of considering stakeholder diversity in soft skills assessment, as it significantly impacts team cohesion and communication.

Continuous feedback mechanisms are vital for enabling real-time adjustments and fostering ongoing learning within projects. The importance of dynamic feedback loops in maintaining relevance and competitiveness is emphasized, ensuring that skills and contextual understanding are constantly updated and refined. This continuous improvement process is crucial for organizations operating in rapidly evolving landscapes.

The integration of multidimensional soft skills ensures that employees are well-rounded and capable of addressing a broad spectrum of challenges. Effective communication within a team, for instance, varies significantly depending on the project's specific requirements and the overall cultural context. Our model ensures that employees are equipped to handle diverse tasks and challenges, promoting alignment, flexibility, and adaptability.

The development and application of the 3SM are grounded in several key organizational theories that collectively provide a robust theoretical foundation. Each of these theories contributes uniquely to understanding and enhancing the dynamic and multifaceted nature of soft skills in project management.

Contingency Theory closely aligns with the Contextual Embeddedness (CE) and Project-Specific Adaptability (PSA) components of the Shapeshifter Model. This theory posits that there is no single best way to manage an organization; instead, the optimal approach depends on various internal and external factors. In project management, this translates to the necessity for project managers to tailor their soft skills and management strategies to the unique cultural, social, political, and institutional circumstances of each project. By applying Contingency Theory, we underscore the critical importance of situational awareness and adaptability in the effective deployment of soft skills.

The Resource-Based View (RBV) emphasizes the strategic importance of internal resources in achieving a competitive advantage. Within the Shapeshifter Model, soft skills are seen as valuable resources that can significantly enhance a project manager's effectiveness. The concept of Multidimensional Skills Integration (MSI) within our model highlights the need for project managers to develop a comprehensive set of soft skills across various domains. By framing soft skills as strategic resources, RBV supports our assertion that cultivating and integrating these skills can lead to superior project outcomes and overall organizational performance.

The Dynamic Capabilities Framework focuses on an organization's ability to integrate, build, and reconfigure internal and external competencies to address rapidly changing environments. This framework is particularly relevant to the Shapeshifter Model's emphasis on Dynamic Feedback Loops (DFL) and Situational Responsiveness (SR). Project environments are inherently dynamic, requiring continuous learning and adaptation. By leveraging dynamic capabilities, project managers can refine and adjust their soft skills in real time, ensuring they remain effective and relevant in response to evolving project demands and stakeholder feedback.

Stakeholder Theory considers the interests and influences of all stakeholders involved in or affected by an organization. The Shapeshifter Model's focus on Stakeholders' Diversity (SD) aligns well with this theory. Effective project management requires recognizing and managing the diverse expectations, communication styles, needs, and behaviors of various stakeholders. Stakeholder Theory supports our model by highlighting the importance of understanding and navigating stakeholder diversity to achieve successful collaboration and project outcomes. By incorporating the principles of Stakeholder Theory, the Shapeshifter Model ensures that project managers are equipped to handle the complexities of stakeholder interactions in diverse project environments.

Complexity Theory addresses the behavior of complex systems and how they adapt and evolve. This theory is relevant to understanding the multilayered and dynamic nature of soft skills in project management, as outlined in our model. The interaction between individual differences, interpersonal interactions, and evolving contextual and project-specific factors creates a complex environment where soft skills are continually activated and redefined. By applying Complexity Theory, we emphasize the need for project managers to navigate this complexity through continuous learning, adaptation, and situational responsiveness.

By grounding the 3SM in these theories, we provide a robust framework that not only captures the dynamic and multifaceted nature of soft skills in project management but also offers practical insights for enhancing project success through effective soft skill integration and adaptation.

7 Conclusions

This research addresses the inherent limitations of traditional, static approaches to assessing soft skills by proposing a novel framework that incorporates contextual variables and stakeholders' diversity into the evaluation process. Our findings suggest that a dynamic and context-aware assessment model provides a more comprehensive and accurate representation of an individual's soft skills.

The proposed model allows for a nuanced understanding of soft skills by recognizing the influence of varying contexts on their manifestation and effectiveness. By integrating contextual variables, the assessment process becomes more reflective of real-world scenarios where soft skills are applied, thereby enhancing the validity and reliability of the evaluation.

Furthermore, this research contributes to the field by offering a methodological advancement that bridges the gap between theoretical constructs of soft skills and their practical assessment. The use of contextual variables ensures that the assessment is adaptable and sensitive to the specificities of different environments, leading to more personalized and relevant evaluations.

The 3SM model offers a structured approach to understanding and integrating the elements of the provided framework. By focusing on multidimensional skill integration, contextual embeddedness, project-specific adaptability, stakeholders' diversity and dynamic feedback loops, it provides a robust and flexible system for projects to thrive in complex and dynamic environments.

8 Implications

Our framework represents a significant step forward in the assessment of soft skills, offering a robust, flexible, and contextually informed approach that better captures the complexities of human competencies. This advancement holds the potential to improve educational practices, inform policy decisions, and enhance workforce development strategies.

Implementing the 3SM involves several practical steps. Organizations should regularly update their training programs to include emerging skills. This proactive approach to skill development ensures that project managers remain capable of meeting new challenges.

Environmental scanning systems should be established to monitor external factors that influence the project. By staying informed about political, social, and cultural shifts, project specificities organizations can make strategic adjustments that enhance their resilience and adaptability.

Lastly, organizations should develop robust feedback mechanisms to gather insights from stakeholders continuously. These insights should be used to refine soft skills, fostering a culture of continuous improvement that keeps the organization dynamic and responsive to change.

Acknowledgments. A heartfelt thank you to Camilla Capogreco, Marco Cerri, and Emanuele Patti for their invaluable contributions to the definition and refinement of the conceptual framework.

Limitations. While our framework aims to integrate contextual, project-specific, situation and stakeholders' diversity variables, the theoretical nature of this study means that we have not yet explored how these variables interact with soft skills in diverse project settings. Empirical research is necessary to understand the nuances and complexities of these interactions.

Further Avenues for Research. While our framework accounts for contextual variables and stakeholders' diversity, future studies could refine and adapt this model for specific industries or types of projects. This would involve testing and validating the 3SM in diverse project settings to enhance its applicability and robustness.

References

1. Alvarenga JC et al (2020) The project manager core competencies to project success. Int J Manag Proj Bus 13(2):277–292
2. Brunet M (2022) On the relevance of theory and practice in project studies. Int J Proj Manag 40(1):22–24
3. Buchanan DA (1991) Vulnerability and agenda: context and process in project management. Br J Manag 2(3):121–132
4. Burke R, Barron S (2014) Project management leadership: building creative teams. Wiley
5. Cimatti B (2016) Definition, development, assessment of soft skills and their role for the quality of organizations and enterprises. Int J Qual Res 10(1):97
6. Creasy T, Anantatmula VS (2013) From every direction--how personality traits and dimensions of project managers can conceptually affect project success. Proj Manag J 44(6):36–51
7. Dulewicz V, Higgs M (2005) Assessing leadership styles and organizational context. J Manag Psychol 20(2):105–123
8. Goleman D (1999) What makes a leader. IEEE Eng Manag Rev 27:4–11
9. Heckman JJ, Kautz T (2012) Hard evidence on soft skills. Labour Econ 19(4):451–464
10. Ibrahim R et al (2017) The effect of soft skills and training methodology on employee performance. Eur J Train Dev 41(4):388–406
11. Indelicato G (2012) Book review: the complete project manager: integrating people, organizational, and technical skills. Proj Manag J 43(6):113
12. Joslin R, Müller R (2015) Relationships between a project management methodology and project success in different project governance contexts. Int J Proj Manag 33(6):1377–1392
13. Kerzner H (2017) Project Management: a systems approach to planning, scheduling, and controlling. Wiley
14. Laker DR, Powell JL (2011) The differences between hard and soft skills and their relative impact on training transfer. Hum Resour Dev Q 22(1):111–122
15. De Lima LF, Silva SQ (2020) Emotional intelligence and success of project management: the mediating effect of interpersonal skills. Int J Proj Organ Manag 12(1):54–73
16. Manning S (2008) Embedding projects in multiple contexts–a structuration perspective. Int J Proj Manag 26(1):30–37
17. Maqbool R, Sudong Y, Manzoor N, Rashid Y (2017) The impact of emotional intelligence, project managers' competencies, and transformational leadership on project success: an empirical perspective. Proj Manag J 48(3):58–75
18. Marin-Zapata SI, Román-Calderón JP, Robledo-Ardila C, Jaramillo-Serna MA (2022) Soft skills, do we know what we are talking about? Rev Manag Sci 16(4):969–1000
19. Martens ML, De Carvalho MM (2017) Key factors of sustainability in project management context: a survey exploring the project managers' perspective. Int J Proj Manag 35(6):1084–1102
20. Martinsuo M (2013) Project portfolio management in practice and in context. Int J Proj Manag 31(6):794–803
21. Martinsuo M, Ahola T (2022) Multi-project management in inter-organizational contexts. Int J Proj Manag 40(7):813–826
22. Muzio E, Fisher DJ, Thomas ER, Peters V (2007) Soft skills quantification (SSQ) for project manager competencies. Proj Manag J 38(2):30–38
23. Müller R, Turner JR (2010) Leadership competency profiles of successful project managers. Int J Proj Manag 28(5):437–448
24. Pellegrinelli S, Partington D, Hemingway C, Mohdzain Z, Shah M (2007) The importance of context in programme management: an empirical review of programme practices. Int J Proj Manag 25(1):41–55

25. Sunindijo RY, Zou PX (2013) The roles of emotional intelligence, interpersonal skill, and transformational leadership on improving construction safety performance. Construct. Econ. Build. 13(3):97–113
26. Turner JR, Müller R (2005) The project manager's leadership style as a success factor on projects: a literature review. Proj Manag J 36(2):49–61
27. Zhang F, Zuo J, Zillante G (2013) Identification and evaluation of the key social competencies for Chinese construction project managers. Int J Proj Manag 31(5):748–759
28. Ziek P, Anderson JD (2015) Communication, dialogue and project management. Int J Manag Proj Bus 8(4):788–803
29. vom Brocke J, Zelt S, Schmiedel T (2016) On the role of context in business process management. Int J Inf Manag 36(3):486–495

Meta-Soft Skills Development in Megaproject Management: An Educational Response to "Grand Challenges"

Roberta Virtuani(✉), Barbara Barabaschi, and Franca Cantoni

Università Cattolica del Sacro Cuore, Milan, Italy
roberta.virtuani@unicatt.it

Abstract. By conceptualizing megaprojects as systems of interdependent subprojects, we argue that beyond the conventional skills required for project management, specific meta-skills are crucial for supporting integration and coordination activities. Meta-soft skills are higher-order, cross-functional competencies that enable project managers to synthesize and apply traditional soft skills in more complex and adaptive ways. Rather than being task-specific, these skills facilitate a holistic approach to managing the dynamic and interconnected elements of megaprojects. This study, drawing on an extensive review of project management literature, identifies key soft skills essential for project managers and provides a critical reflection on the meta-skills required to enhance collaboration, integration, and innovation within megaproject environments. Moreover, we propose an educational model tailored to cultivate these meta-soft skills, bridging theoretical knowledge with practical application. By positioning meta-soft skills as integral to megaproject leadership and performance, this research provides novel insights and actionable strategies to address the multifaceted challenges of megaprojects, ultimately contributing to more resilient and adaptive project outcomes in the face of global complexity and uncertainty.

Keywords: megaprojects · meta-soft skills · grand challenge

1 Introduction

Megaprojects, characterized as large-scale, complex, and singular capital investments in both public and private sectors, represent highly risky ventures that are notoriously difficult to manage and frequently fail to meet their original objectives (Denicol et al. 2020). These projects are typically structured as temporary standalone organizations and are classified as megaprojects when their total capital cost exceeds US$1 billion (Flyvbjerg 2017). The critical dimensions contributing to the challenges of managing megaprojects include their scale, uncertainty, complexity, urgency, and institutional structure (Pitsis et al. 2018). These factors are central to much of the existing research aimed at understanding the high frequency of failure in megaprojects. The megaproject is usually decomposed into many smaller inter-related projects and organized as a program. A large organization—the client, prime contractor, and/or delivery partner—is established

E. Favari et al. (Eds.): MeRIT 2024, LNCE 600, pp. 20–34, 2025.
https://doi.org/10.1007/978-3-031-92797-3_2

to coordinate and integrate the efforts of numerous subgroups and suppliers involved in project activities (Davies and Mackenzie 2014; Davies et al. 2009; Söderlund et al. 2017). A notable example of a successful megaproject is the London Olympics Construction Programme, which was both large and highly complex. The programme encompassed the construction of the majority of venues for the 2012 London Olympics, comprising over 70 individual projects, each interdependent on shared services and site logistics.

The division of the megaproject into discrete, interconnected projects proved to be an effective strategy for managing its structural complexity (Damayanti et al. 2020) This segmentation not only allowed for more manageable sub-units but also highlighted the critical role of integration between these projects. Conceptualizing megaprojects as systems composed of interdependent sub-projects offers a powerful analytical framework, enabling a deeper synthesis of insights from both megaproject and traditional project management literature (Davies and Mackenzie 2014) This systems-based approach underscores the inherent complexity of megaprojects, where the success of the overall venture depends on the seamless integration, coordination and alignment of its constituent parts.

Our research makes a distinct contribution by extending the current understanding of soft skills in project management to this more complex, interconnected megaproject domain. By recognizing the unique dynamics and the critical role that soft skills—such as communication, leadership, and integration—play in orchestrating these interdependent sub-projects, we shed light on the necessity of a more nuanced approach to megaproject management. This framework emphasizes that soft skills are not only important in managing individual projects but are also essential for navigating the complex interactions and dependencies that define megaproject environments.

By rigorously analyzing the analogies and distinctions between project and megaproject management, particularly concerning soft skills, we address the unique complexities inherent in managing large-scale ventures.

In doing so, it becomes evident that while strategic and organizational dimensions of megaproject management have been well-studied, the individual level—especially in relation to soft skills—has received far less attention. Yet, this individual level is critical, as it forms the bedrock upon which broader organizational and strategic competencies are developed. Our research seeks to address this gap by highlighting the essential role that individual competencies play in driving megaproject success. By focusing on the individual perspective we aim to enrich the broader understanding of how soft skills contribute to the effective management of complex, large-scale projects.

A key mechanism for addressing this gap is education. The development of soft skills through targeted educational initiatives is increasingly essential, especially in an era of constant organizational change and "grand challenges" (Ika et al. 2024). Education provides a crucial link between theory and practice, enabling executives and managers to adapt more effectively to dynamic environments. By cultivating a culture that prioritizes soft skills development, organizations can better align their workforce with the demands of complex megaprojects. Our research contributes to this by identifying the most critical soft skills for project and megaproject management and proposing an educational framework that develops these competencies to prepare employees, managers, and leaders for future challenges.

Traditional educational models, which have largely relied on passive learning methods such as frontal lessons, are no longer sufficient to meet the evolving needs of megaproject management. Advancements in technology, particularly in the fields of augmented reality and artificial intelligence, offer new, more effective methods for training and developing soft skills. These innovative approaches enable learners to engage with complex scenarios and challenges in real time, fostering the adaptability and problem-solving abilities necessary for success in megaprojects.

The chapter is organized as follows: after the introduction we present the literature review dealing with the subject of soft skills in project management to identify the five most cited soft skills to develop for the management of a project. The following Sect. 2 deepen the main meta soft skills for megaprojects at the integration and coordination organizational level. These are the skills that characterize megaprojects executives to achieve the venture results. In Sect. 3 the future of teaching soft skills to megaproject managers at integration and operational level is analysed. New technologies can be of great effectiveness to train and develop soft skills to boost the megaproject success. As the world of megaprojects becomes increasingly complex and interconnected, so too must be the methods we use to train the leaders responsible for their success. Conclusions and future research avenues end the chapter.

2 Literature Review

Traditionally, PM was largely synonymous with technical prowess and proficiency in the intricacies of project methodologies. The emphasis was placed squarely on the mastery of tools, processes, and techniques, often overlooking the human element (Alvarenga et al. 2020). However, the recognition that projects are fundamentally social endeavors, shaped by the collaboration and interaction of diverse individuals (Brunet 2022; De Lima and Silva 2020) in ever-changing organizational landscapes (Martens and De Carvalho 2017; Martinsuo 2013) has given way to a more holistic understanding that integrates technical competence with a rich tapestry of soft skills (Zhang et al. 2013). Moreover, in the intricate landscape of PM (Locatelli et al. 2023), soft skills serve as the glue that binds together diverse stakeholders, facilitating collaboration conflict resolution and consensus-building (Magano et al. 2020). Effective communication skills enable project managers to convey objectives, expectations, and feedback transparently, fostering alignment and understanding among team members (Elkbuli et al. 2024). Furthermore, adept interpersonal skills empower project managers to navigate the complexities of team dynamics, harnessing individual strengths and mitigating conflicts to maintain cohesion and momentum (Ojiako et al. 2011). Additionally, soft skills are instrumental in leadership within project contexts (Anderson 2010). A project manager's capacity to inspire and motivate team members, fostering a culture of collaboration, resilience, and accountability, enhance team engagement and productivity, ultimately driving project success.

Recognizing and cultivating these skills is paramount for project managers seeking to optimize performance, foster innovation, and deliver value in today's dynamic and interconnected business environment (Hefley and Bottion 2021).

Consequently, there has been a growing awareness that delineating the ideal profile for a project manager who can effectively contribute to project success or seamlessly integrate within a team dynamic involves a primary challenge that extends beyond merely assessing technical proficiency. Rather, it requires a comprehensive understanding of their repertoire of soft skills.

One of the primary objectives of this research was to identify the predominant soft skills documented in the existing literature concerning PM. The initial phase of the study involved a systematic approach to keyword selection, encompassing a comprehensive range of terminology related to soft skills.

The intentional selection of Scopus as the primary database, coupled with the inclusion of document types such as articles and early access materials, was a methodological decision aimed at facilitating a thorough and exhaustive review process. To further enhance the comprehensiveness of the review, inclusion criteria were strategically formulated to concentrate on specific Scopus categories pertinent to the fields of management, education, business, and engineering. By doing so, the interdisciplinary nature of the research endeavor was accentuated, allowing for a holistic examination of relevant literature spanning multiple domains. Moreover, the establishment of language criteria, with a preference for English-language publications, was intended to promote inclusivity within the global academic community. We applied the following string: *(("project management" AND ("soft skill*" OR "transversal skill*" OR "personal skill*" OR "interpersonal skill*")).*

The abstracts of the papers retrieved through the search string undergone a screening process to determine their eligibility. These abstracts were independently reviewed by the authors and filtered based on their relevance to the research objectives. Papers that meet the criteria were read in full, and the soft skills mentioned within them were mapped accordingly. Additionally, a representation of these findings were created using VOSviewer.

As shown in Fig. 1, a total of 182 papers were initially identified. Upon a detailed review of the abstracts, it was determined that only 54 papers were directly relevant to the specific objectives outlined in our research framework. A comprehensive analytical examination of these selected papers enabled the systematic identification and mapping of 82 distinct soft skills within the context of PM.

Using VosViewer, Fig. 1 provides a visualization of the 82 identified soft skills.

The mapping of soft skills provides a comprehensive framework for understanding how different competencies are prioritized in the literature, offering critical insights for targeted skill development in professional contexts. The scientific mapping and literature review identified several soft skills that are frequently cited as essential for effective project and megaproject management.

Communication skills emerge as a fundamental component, as clear and effective communication is essential for fostering collaboration, preventing miscommunication, and facilitating cohesive team dynamics. It ensures that all stakeholders remain aligned, which is crucial for maintaining project momentum. Another key skill is leadership, which manifests in various forms, including transformational, inspirational, and external leadership. These leadership styles are critical for guiding teams, driving organizational change, and motivating individuals to achieve project objectives, particularly in the complex environments of projects and megaprojects.

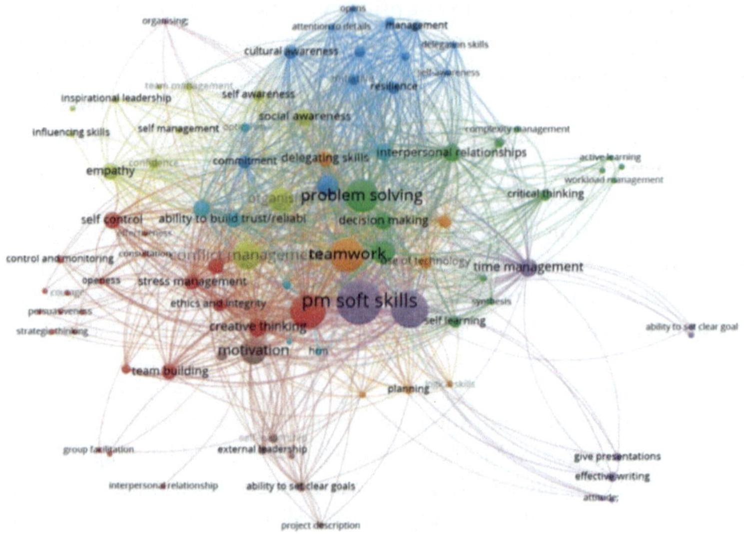

Fig. 1. Soft Skills found with the literature review

Time management is consistently recognized as a crucial capability, encompassing the ability to organize tasks efficiently, set realistic goals, and meet deadlines. This skill is vital for both individual and collective performance, ensuring that projects are delivered on time and within scope. Equally important is the ability to resolve conflicts and negotiate effectively. The capacity to manage conflicts, find diplomatic compromises, and resolve crises is invaluable, particularly in high-pressure situations. Successful conflict resolution helps maintain smooth operations and minimizes disruptions throughout the project lifecycle.

Finally, resourcefulness—characterized by creativity, problem-solving, and innovative thinking—is essential for overcoming unforeseen challenges and fostering continuous improvement within project environments. Resourcefulness enhances adaptability and promotes the development of innovative solutions to complex problems.

3 Megaprojects Meta-Soft Skills

According to this definition, we contend that in the management of megaprojects, in addition to the conventional skills required for project management, specific meta-skills are essential to support integration and coordination activities. Meta-soft skills refer to higher-order, cross-functional competencies that enable project managers to synthesize and apply various traditional soft skills in more complex and adaptive ways. These skills are not task-specific but rather facilitate a more holistic approach to managing the dynamic, interconnected elements of megaprojects.

Meta-skills are critical for navigating the complex interdependencies inherent in megaprojects, ensuring that project leaders can effectively integrate diverse sub-projects and align them with the overarching goals. By enabling seamless coordination across multiple organizational layers and disciplines, these meta-skills help to mitigate risks, foster collaboration, and enhance the capacity to adapt to unforeseen challenges. As such, they are indispensable for ensuring the overall success of megaprojects, where complexity and uncertainty often exceed those found in traditional project management contexts.

3.1 Leadership

Leadership in the context of megaproject management extends beyond the traditional roles of directing teams and overseeing task execution. Given the scale, complexity, and interdependencies inherent in megaprojects, leadership requires the application of meta-soft skills that allow for adaptive and dynamic management. These meta-skills, such as systems thinking, emotional intelligence, and strategic foresight, enable leaders to navigate the multifaceted nature of megaprojects and ensure the successful integration and alignment of multiple sub-projects (Muller et al. 2022).

At the heart of effective megaproject leadership is the ability to synthesize traditional soft skills—such as communication, decision-making, and conflict resolution—into a more comprehensive, higher-order set of competencies. Leaders must be able to recognize and manage the intricate interrelationships between various project components, often across different organizational, cultural, and disciplinary boundaries. This requires not only technical and logistical expertise but also the capacity to foster collaboration, build trust among stakeholders, and adapt to rapidly changing conditions (Merrow and Nandurdikar 2018).

In particular, megaproject leaders must demonstrate the ability to coordinate diverse teams, ensuring that all actors are aligned with the overarching project goals while maintaining flexibility to adjust strategies in response to emerging challenges. Meta-leadership in this context includes the skill to manage ambiguity, foresee potential disruptions, and integrate different perspectives into coherent, strategic decision-making processes (Northouse 2021).

Furthermore, the leadership required for megaprojects involves balancing short-term operational demands with long-term strategic vision. The leader's role is to create an environment that promotes continuous learning and innovation while ensuring that project goals are met in a timely and cost-effective manner. This requires leveraging emotional intelligence to inspire and motivate teams, particularly in high-stress, high-stakes environments typical of megaprojects (Whyte et al. 2022).

By applying meta-soft skills, leaders can navigate the complexity and uncertainty that define megaproject environments. These skills help mitigate risks by ensuring that the various sub-projects are not only aligned with the broader project objectives but also capable of adapting to changes in scope, resources, and external conditions. Thus, leadership in megaprojects is a dynamic and multifaceted role, one that demands a balance between operational oversight and strategic vision, underpinned by a strong foundation of meta-soft skills that go beyond conventional project management practices (Virtuani et al. 2022)

3.2 Team Management

In the context of megaprojects, team management transcends traditional project coordination and operational oversight, requiring the application of meta-skills that support effective integration, collaboration, and adaptability across diverse teams. The complexity of megaprojects—characterized by their scale, multiple interdependent sub-projects, and multidisciplinary nature—demands a more holistic approach to managing teams, one that integrates both technical and interpersonal dimensions.

Meta-skills, such as emotional intelligence, cultural awareness, and systems thinking, are crucial for managing the diverse human capital involved in megaprojects. These higher-order competencies enable project managers to navigate the complexities of coordinating teams that may operate across different geographic locations, organizational structures, and cultural backgrounds. Effective team management in this context requires a dynamic and adaptive leadership style, where team members are empowered to contribute their expertise while being aligned with the broader project objectives (Biesenthal et al. 2017; Struggles and Heindrick 2015).

A critical component of team management in megaprojects is the ability to foster collaboration and integration among diverse stakeholders. Project managers must not only coordinate tasks but also ensure that team members from various disciplines and backgrounds work together cohesively. This involves creating an environment of trust, open communication, and mutual respect, where conflicts are resolved constructively, and differences in professional and cultural perspectives are leveraged to enhance innovation and problem-solving. Meta-skills, such as conflict resolution and negotiation, play a pivotal role in ensuring that these interactions are productive and aligned with the project's strategic goals (Barabaschi et al. 2020).

Moreover, effective team management in megaprojects requires the ability to maintain flexibility and adaptability in the face of evolving project conditions. The dynamic nature of megaprojects often necessitates rapid adjustments in team structure, resource allocation, and task prioritization. Managers must be able to reconfigure teams and processes in response to new challenges, all while maintaining alignment with the project's overarching objectives. Meta-skills such as strategic foresight and decision-making under uncertainty are essential in guiding these adjustments and ensuring that teams remain resilient and focused on delivering project outcomes.

In addition, managing megaproject teams requires a deep understanding of the interdependencies between different team functions and roles. Project managers must be able to anticipate how changes in one sub-project or discipline may impact others and adjust team workflows accordingly. This systems-based approach to team management ensures that the overall integration of the megaproject is maintained, reducing the risk of misalignment or delays due to miscommunication or isolated decision-making.

Ultimately, the success of team management in megaprojects hinges on the project manager's ability to apply these meta-skills to create an integrated, adaptive, and collaborative working environment. By fostering a culture of shared responsibility, innovation, and continuous learning, project managers can enhance team performance and ensure that the complex interdependencies of megaprojects are effectively managed. This holistic approach to team management not only improves operational efficiency but also contributes to the overall success of the megaproject in achieving its strategic goals.

3.3 Stakeholder Engagement

Stakeholder engagement in megaproject management transcends traditional communication and stakeholder management practices, requiring the application of meta-skills that foster trust, collaboration, and adaptability among diverse groups. Megaprojects are inherently complex, involving a wide range of internal and external stakeholders, including governments, regulatory bodies, contractors, local communities, and private organizations. Each of these stakeholders brings unique interests, objectives, and cultural backgrounds, which can either support or hinder the project's progress. Therefore, successful stakeholder engagement demands a holistic, systems-based approach that integrates both technical and interpersonal competencies to align these varied interests with the overarching goals of the megaproject (Liu et al. 2018; Mok et al. 2015).

Meta-skills, such as emotional intelligence, cultural awareness, and negotiation, are fundamental to engaging stakeholders effectively. These higher-order competencies enable project managers to navigate the political, social, and organizational dynamics that shape stakeholder relationships. Emotional intelligence, for example, allows project leaders to understand and respond to the concerns of different stakeholders, fostering mutual respect and trust. Cultural awareness is critical in managing stakeholders from diverse backgrounds, ensuring that engagement strategies are inclusive and respectful of different values and practices.

A key element of stakeholder engagement in megaprojects is the ability to maintain open lines of communication, where transparency and trust-building are prioritized (Ninan et al. 2019). Project managers must establish communication frameworks that allow for continuous feedback, ensuring that stakeholders are kept informed and their concerns are addressed in a timely manner. This process not only helps to prevent conflicts but also facilitates proactive problem-solving. Conflict resolution and negotiation meta-skills are particularly valuable here, as they enable project leaders to mediate disputes and align stakeholder expectations with the project's objectives.

Flexibility and adaptability are also essential for successful stakeholder engagement. As megaprojects evolve, stakeholders' interests and external conditions may shift, requiring project managers to adjust engagement strategies accordingly. Meta-skills such as strategic foresight and decision-making under uncertainty allow project managers to anticipate changes and adapt their approaches to maintain stakeholder alignment. By leveraging these skills, project leaders can mitigate potential risks associated with stakeholder dissatisfaction, thereby enhancing the project's resilience and capacity to navigate challenges.

Additionally, effective stakeholder engagement in megaprojects involves recognizing the interdependencies between different stakeholder groups. Project managers must understand how the actions, needs, and influence of one stakeholder can impact others, and adjust their engagement strategies to ensure harmonious relationships across the project ecosystem. A systems-based perspective, supported by meta-skills like systems thinking and integrative leadership, enables project leaders to manage these interdependencies, ensuring that stakeholder interactions are productive and contribute to the successful integration and alignment of the project's goals.

Ultimately, stakeholder engagement in megaproject management is about creating an inclusive, collaborative environment where diverse interests are aligned with the project's strategic objectives. By applying meta-skills that facilitate communication, negotiation, and adaptability, project managers can build strong, trusting relationships with stakeholders, mitigating risks and enhancing the overall success of the megaproject.

3.4 System Thinking

In the context of megaprojects, systems thinking involves recognizing the project as a complex, adaptive system where individual elements—such as teams, resources, and external stakeholders—are interconnected and interdependent. A change in one part of the system, such as delays in a specific sub-project or shifts in stakeholder priorities, can ripple through the entire project, affecting timelines, budgets, and overall outcomes. By adopting a systems-based perspective, megaproject managers can move beyond a linear, cause-and-effect approach to decision-making, enabling them to identify the broader implications of local actions and anticipate emergent behaviors within the system.

One of the key applications of systems thinking in megaproject management is the identification and management of feedback loops. These loops can be either reinforcing or balancing, and they significantly influence project dynamics. Reinforcing loops, for instance, can accelerate project outcomes—both positively and negatively—while balancing loops help stabilize processes. Systems thinking enables managers to recognize and manipulate these loops to mitigate risks or enhance positive outcomes. For example, a delay in one sub-project might create a reinforcing negative loop, causing further delays and cost overruns. Recognizing this early allows the manager to intervene with appropriate adjustments, potentially stabilizing the project (Sankaran et al. 2010).

Additionally, systems thinking equips megaproject managers with the ability to model and simulate the potential outcomes of various scenarios. By visualizing the complex relationships between different project components, systems thinking allows managers to test different strategies and predict their impact on the project as a whole. This is particularly useful in the dynamic environment of megaprojects, where unexpected changes are common. Scenario modeling enables managers to develop adaptive strategies that account for uncertainty, improving their ability to navigate unforeseen challenges (Shams Esfandabadi et al. 2022).

Another critical aspect of systems thinking in megaprojects is its emphasis on integration. Since megaprojects are typically composed of numerous sub-projects that must work together seamlessly to achieve the overall project goals, systems thinking encourages a focus on the coordination and alignment of these elements. By understanding how different sub-projects are interconnected, managers can ensure that resources, information, and efforts are harmonized across the project. This reduces inefficiencies, prevents silos from forming, and ensures that the project operates as a cohesive whole rather than as disjointed parts.

Moreover, systems thinking supports a deeper understanding of the external environment in which the megaproject operates. Megaprojects are often subject to external influences such as regulatory changes, economic fluctuations, and societal pressures. Systems thinking allows project managers to map these external factors and understand how they interact with internal project dynamics. By doing so, managers can anticipate potential disruptions and develop strategies to mitigate their impact, ensuring greater project resilience.

3.5 Ethics

Ethics in megaprojects is a fundamental meta-skill that involves guiding decision-making and behavior in alignment with broader societal, environmental, and stakeholder responsibilities. In the complex and interconnected nature of megaprojects, ethical considerations transcend traditional notions of compliance and governance, encompassing a deeper responsibility to manage the moral implications of actions across all phases of the project. Ethical leadership in this context requires a commitment to transparency, accountability, and fairness, ensuring that the diverse and often conflicting interests of stakeholders—ranging from governments and investors to local communities—are balanced and addressed equitably (Jonasson and Ingason 2013).

The dynamic and uncertain environments in which megaprojects operate often present ethical dilemmas, such as the allocation of resources, the environmental impact of large-scale construction, or the displacement of communities. Meta-skills, such as empathy, cultural awareness, and systems thinking, are essential for navigating these dilemmas, as they allow project leaders to anticipate and mitigate ethical risks while fostering sustainable and socially responsible outcomes. These meta-skills enable a holistic perspective on how individual decisions affect the broader project ecosystem, encouraging project managers to weigh long-term consequences alongside short-term goals.

Ethics in megaprojects also involves a commitment to sustainability, ensuring that the project contributes to long-term environmental and social well-being without compromising future generations. This includes integrating sustainable practices into project planning, execution, and governance, ensuring that project outcomes are not achieved at the expense of ecological degradation or social inequity. Ultimately, ethics in megaproject management is about fostering a culture of integrity and responsibility, where leaders and teams are accountable not only to project objectives but also to the broader public good.

4 The Future of Teaching Soft Skills to Megaproject Managers

As the scale and complexity of global megaprojects continue to expand, the ability to cultivate advanced meta-soft skills in megaproject managers has become paramount. The challenge lies in how these can be taught effectively to a new generation of leaders, particularly those from Generation Z, whose learning preferences and cognitive processes differ markedly from prior cohorts (Chillakuri and Mahanandia 2018). Generation Z, with its deep digital fluency and preference for highly interactive, personalized learning environments, requires innovative approaches that move beyond passive knowledge transmission (Cilliers 2017; Dwivedi et al. 2019). The integration of immersive learning technologies, such as augmented reality (AR) and virtual reality (VR), into soft skills training for megaproject managers represents a significant leap forward (Safikhani et al. 2022; Tender et al. 2024). In this context, AR and VR offer the opportunity to place learners in hyper-realistic simulations of megaproject environments, where they can practice leadership, communication, and team dynamics in high-stakes, pressure-laden scenarios. For instance, a manager might navigate a virtual stakeholder meeting involving government officials, contractors, and local community leaders, where competing interests and time constraints force real-time decision-making. This immersive environment not only engages learners but also replicates the emotional and psychological pressures they would face in actual megaprojects, allowing for a deeper, more visceral understanding of how to apply soft skills in complex interpersonal interactions (Cantoni et al. 2024).

In parallel, the use of AI-driven adaptive learning platforms can revolutionize how soft skills are developed in megaproject management. These platforms, equipped with sophisticated algorithms, can continuously monitor and assess each learner's progress, dynamically adjusting the difficulty of exercises and offering personalized feedback based on individual learning patterns (Dutta et al. 2024; Ezzaim et al. 2024). This kind of real-time adaptability ensures that the training is responsive to the specific needs of each participant, maximizing skill acquisition and retention. For megaproject managers, such platforms can simulate evolving project conditions—ranging from sudden resource shortages to unexpected stakeholder opposition—thereby training participants to apply negotiation and conflict resolution skills under fluctuating conditions (Sarnato et al. 2024). Moreover, AI's ability to analyze behavioral data can provide insights into a learner's decision-making tendencies, offering targeted interventions to address blind spots in leadership style or communication strategies.

A further innovation in teaching soft skills to megaproject managers lies in the concept of "learning ecosystems" (Kowch 2018) In contrast to traditional siloed learning environments, learning ecosystems emphasize an interconnected web of resources, tools, and peer interactions that mirror the complex, multi-stakeholder landscape of megaprojects. By employing blockchain-powered credentialing systems, learners can gain verifiable, micro-credentials for mastering specific soft skills, such as cross-cultural communication or stakeholder engagement. These credentials can be shared across an ecosystem of collaborators, including universities, industry organizations, and project stakeholders, ensuring that soft skill proficiency is recognized and valued across the megaproject lifecycle (Tran et al. 2023). Such ecosystems promote continuous, lifelong learning, where managers can access bite-sized learning modules on demand, reflect on previous project experiences, and apply new strategies as project conditions evolve.

Gamification also remains a vital element in the learning process but can be pushed further through the development of "meta-project simulations" that integrate the entirety of the megaproject experience. Rather than isolated role-playing exercises, meta-project simulations would allow learners to engage with long-term, scenario-based games where the outcome depends on cumulative decision-making across multiple project phases (Hass et al. 2021; Jayawardena 2021). Within these simulations, soft skills such as emotional intelligence, leadership adaptability, and stakeholder negotiation would be tested in tandem with technical and financial decisions, offering a holistic view of how human dynamics impact megaproject outcomes. For example, a team of learners could be tasked with managing a large infrastructure project over several months, with AI-generated scenarios that introduce unpredictable challenges like political instability or workforce strikes. This immersive gamified experience forces learners to engage continuously with their soft skills in real-world timeframes, providing a richer, more integrated approach to learning.

Incorporating quantum computing-powered simulations and neuroscientific feedback mechanisms could further push the boundaries of soft skills education. By integrating neural feedback into soft skills training, learners can receive real-time insights into their emotional and cognitive states during high-stress simulations. This data could be used to fine-tune leadership approaches, improve emotional regulation, and optimize communication strategies. Quantum computing, with its unparalleled processing power, could simulate highly complex megaproject scenarios with thousands of variables, allowing managers to engage with unprecedented levels of project realism, testing their soft skills against multiple intersecting crises in ways that classical computing cannot replicate.

In conclusion, the future of teaching soft skills to megaproject managers lies at the intersection of technology, personalized learning, and immersive experiences (Shelley 2015). The integration of cutting-edge tools such as AR, VR, AI-driven adaptive learning, and blockchain-based ecosystems provides an unprecedented opportunity to transform how we cultivate the interpersonal competencies necessary for leading megaprojects. As the world of megaprojects becomes increasingly complex and interconnected, so too must the methods we use to train the leaders responsible for their success. By embracing these innovations, we can prepare a new generation of megaproject managers who are not only technically proficient but also masters of the human dynamics that determine project outcomes.

5 Conclusions

This paper has emphasized the critical role of meta-soft skills in the successful management of megaprojects, highlighting their significance in navigating the complexities of large-scale, interconnected projects. By conceptualizing megaprojects as systems composed of interdependent sub-projects, we have shown that the conventional skillset for project management is insufficient in addressing the dynamic and uncertain environments characteristic of megaprojects. Instead, higher-order competencies—referred to as meta-soft skills—are essential for ensuring seamless integration, coordination, and adaptability across organizational layers and diverse stakeholder groups.

Key meta-soft skills enable project managers to operate holistically, synthesizing traditional soft skills in more complex and adaptive ways. These competencies not only enhance the capacity to manage the intricate interdependencies within megaprojects but also foster the resilience needed to address unforeseen challenges, mitigate risks, and sustain collaboration.

Through a proposed educational framework, this study contributes to the growing recognition that the development of meta-soft skills is a priority in preparing megaproject leaders for the grand challenges they face. By advancing the understanding of how these skills integrate with both technical competencies and project management principles, we offer actionable insights for training and education in this field.

6 Future Research Avenues

This study underscores the critical importance of meta-soft skills in megaproject management, yet several areas remain underexplored and warrant further investigation. A key area for future research lies in understanding how these meta-soft skills can be effectively integrated into existing project management training frameworks, particularly for megaproject leaders. The development of higher-order competencies must be explored within the context of current educational practices to identify the most effective methods for embedding these skills into formal and informal training programs.

Emerging technologies present another promising area of inquiry. Future research should investigate the role of innovations such as artificial intelligence, augmented reality, and immersive virtual simulations in enhancing the development of meta-soft skills. These technologies offer the potential to create more interactive, real-time learning environments that mirror the complexities and unpredictability of megaproject scenarios. Understanding how these technologies can be harnessed to develop adaptive and responsive skills in project managers is essential for advancing educational methods in this field.

Furthermore, there is a need to explore more precise methods for assessing and measuring the impact of meta-soft skills on the success of megaprojects. Current research has largely focused on the technical and procedural aspects of project success, while the contribution of soft skills—especially meta-soft skills—remains difficult to quantify. Investigating innovative ways to evaluate how these competencies influence project outcomes will provide a deeper understanding of their practical relevance.

Finally, future research should investigate the long-term benefits of fostering a culture of continuous learning and innovation in megaproject teams. Examining how meta-soft skills contribute to creating such a culture will provide insights into how team performance and adaptability can be enhanced throughout the project life cycle. By addressing these areas, future research can further refine our understanding of the role of meta-soft skills in navigating the complexities of megaproject management.

References

1. Alvarenga JC et al (2020) The project manager core competencies to project success. Int J Manag Proj Bus 13(2):277–292
2. Anderson B (2010) Project leadership and the art of managing relationships. Am Soc Train Dev 64(3):58–63
3. Barabaschi B, Cantoni F, Virtuani R (2020) Rethinking leadership in megaprojects. Results from a preliminary investigation. In: Megaprojects. Multidisciplinarity as the key to successful management. Giappichelli Editore, pp 97–104
4. Biesenthal C, Clegg S, Mahalingam A, Sankaran S (2017) Applying institutional theories to managing megaprojects. Int J Proj Manag 36:43–54. https://doi.org/10.1016/j.ijproman.2017.06.006

5. Brunet M (2022) On the relevance of theory and practice in project studies. Int J Proj Manag 40(1):22–24
6. Cantoni F, Capogreco C, Cerri M, Patti E (2024) Refining the assessment of soft skills in project management. Quaderni del Dipartimento di Scienze Economiche e Sociali, Vita e Pensiero, Milano
7. Chillakuri B, Mahanandia R (2018) Generation Z entering the workforce: the need for sustainable strategies in maximizing their talent. Hum Resour Manag Int Dig 26(4):34–38. https://doi.org/10.1108/HRMID-01-2018-0006
8. Cilliers EJ (2017) The challenge of teaching generation Z. PEOPLE Int J Soc Sci 3(1):188–198
9. Damayanti R, Hartono B, Wijaya A (2020) Complexity, leadership, and megaproject performance: a configuration analysis. J Indus Eng Manag 14(3):570–603
10. Davies A, Mackenzie I (2014) Project complexity and systems integration: constructing the London 2012 Olympics and Paralympics games. Int J Proj Manag 32(5):773–790
11. De Lima LF, Silva SQ (2020) Emotional intelligence and success of project management: the mediating effect of interpersonal skills. Int J Proj Organ Manag 12(1):54–73
12. Denicol J, Davies A, Krystallis I (2020) What are the causes and cures of poor megaproject performance? A systematic literature review and research agenda. Proj Manag J 51(3):328–345. https://doi.org/10.1177/8756972819896113
13. Dwivedi A, Dwivedi P, Bobek S, Sternad Zabukovšek S (2019) Factors affecting students' engagement with online content in blended learning. Kybernetes 48(7):1500–1515. https://doi.org/10.1108/K-10-2018-0559
14. Dutta, S., Ranjan, S., Mishra, S., Sharma, V., Hewage, P., & Iwendi, C.: Enhancing educational adaptability: a review and analysis of ai-driven adaptive learning platforms. In 2024 4th international conference on innovative practices in technology and management (ICIPTM) (pp. 1–5). IEEE (2024)
15. Elkbuli M, Azmy N, Lee CK (2024) Enhancing project manager communication soft skills and risk management practices in Libyan oil and gas construction projects: the role of experience. J Eng Design Technol 23:1017–1038
16. Ezzaim A, Dahbi A, Aqqal A, Haidine A (2024) AI-based learning style detection in adaptive learning systems: a systematic literature review. J Comput Educ 1-39
17. Flyvbjerg B (ed) (2017) The Oxford handbook of megaproject management. Oxford University Press
18. Hass D, Hass A, Joseph M (2021) A preliminary investigation of gamification from the young consumer's perspective. Young People 22(3):413–428. https://doi.org/10.1108/YC-10-2020-1221
19. Hefley WE, Bottion M (2021) Skills of junior project management professionals and project success achieved by them. Int J Inf Syst Proj Manag 9(1):56–75
20. Ika L, Locatelli G, Drouin N (2024) Policy-driven projects: empowering the world to confront grand challenges. Eur Manag J. 42:835–842. https://doi.org/10.1016/j.emj.2024.09.001
21. Jayawardena NS (2021) The e-learning persuasion through gamification: an elaboration likelihood model perspective. Young People 22(3):480–502. https://doi.org/10.1108/YC-08-2020-1201
22. Jonasson HI, Ingason HT (2013) Project ethics. Gower Publishing, Ltd
23. Kowch EG (2018) Designing and leading learning ecosystems: challenges and opportunities. TechTrends 62(2):132–134
24. Liu B, Li Y, Xue B, Li Q, Zou PXW, Li L (2018) Why do individuals engage in collective actions against major construction projects? An empirical analysis based on Chinese data. Int J Proj Manag 36(4):612–626
25. Locatelli G, Ika L, Drouin N, Müller R, Huemann M, Söderlund J, Clegg S (2023) A manifesto for project management research. Eur Manag Rev 20(1):3–17

26. Magano J, Silva C, Figueiredo C, Vitória A, Nogueira T, Pimenta Dinis MA, Generation Z (2020) Fitting project management soft skills competencies—a mixed-method approach. Educ Sci 10(7):187
27. Martens ML, De Carvalho MM (2017) Key factors of sustainability in project management context: a survey exploring the project managers' perspective. Int J Proj Manag 35(6):1084–1102
28. Martinsuo M (2013) Project portfolio management in practice and in context. Int J Proj Manag 31(6):794–803
29. Merrow EW, Nandurdikar N (2018) Leading complex projects: a data-driven approach to mastering the human side of project management. Wiley
30. Mok KY, Shen GQ, Yang J (2015) Stakeholder management studies in mega construction projects: a review and future directions. Int J Proj Manag 33(2):446–457
31. Muller R, Drouin N, Sankaran S (2022) Balanced Leadership. Making the best use of personal and team leadership in projects. Oxford University Press, Oxford
32. Ninan J, Mahalingam A, Clegg S (2019) External stakeholder management strategies and resources in megaprojects: an organizational power perspective. Proj Manag J 50(6):625–640
33. Northouse PG (2021) Leadership: theory and practice, 9th edn. Sage
34. Ojiako U, Ashleigh M, Chipulu M, Maguire S (2011) Learning and teaching challenges in project management. Int J Proj Manag 29(3):268–278
35. Pitsis A, Clegg S, Freeder D, Sankaran S, Burdon S (2018) Megaprojects redefined–complexity vs cost and social imperatives. Int J Manag Proj Bus 11(1):7–34
36. Safikhani S, Keller S, Schweiger G, Pirker J (2022) Immersive virtual reality for extending the potential of building information modeling in architecture, engineering, and construction sector: systematic review. Int J Digit Earth 15(1):503–526
37. Sankaran S, Haslett T, Sheffield J (2010) Systems thinking approaches to address complex issues in project management. In: Paper presented at PMI® Global Congress 2010—Asia Pacific, Melbourne, Victoria, Australia. Project Management Institute, Newtown Square, PA
38. Sarnato AZ, Sari WD, Rahmawati ST, Hidayat R, Patry H (2024) The evolution of E-learning platforms: from U-learning to AI-driven adaptive learning systems. J Soc Sci Util Technol 2(2):289–300
39. Shams Esfandabadi Z, Cottafava D, Corazza L, Scagnelli SD (2022) Sustainability challenges of high-speed railway megaprojects from a systems thinking lens. In: International workshop "a multidisciplinary approach to embrace complexity and sustainability in megaprojects". Springer Nature Switzerland, Cham, pp 331–341
40. Shelley AW (2015) Project management and leadership education facilitated as projects. Int J Manag Proj Bus 8(3):478–490
41. Söderlund J, Sankaran S, Biesenthal C (2017) The past and present of megaprojects. Proj Manag J 48(6):5–16
42. Tender M, Couto JP, Fuller P, Demian P, Chow V, Vaughan A, Santos R (2024) Using virtual reality for occupational safety and health in a UK tunnelling megaproject. In: Proceedings of the institution of civil engineers-management, procurement and law, vol 40, pp 1–8
43. Tran VD, Ata S, Tran TH, Lam DK, Pham HL (2023) Blockchain-powered education: a sustainable approach for secured and connected university systems. Sustain For 15(21)
44. Virtuani R, Barabaschi B, Cantoni F (2022) The human side of megaprojects: leadership style and traits to face growing levels of complexity and uncertainty. In: Sustainability and megaproject development. Routledge, pp 3–24
45. Whyte J, Naderpajouh N, Clegg S, Matous P, Pollack J, Crawford L (2022) Project leadership: a research agenda for a changing world. Proj Leadership Soc 3:100044. https://doi.org/10.1016/j.plas.2022.100044
46. Zhang F, Zuo J, Zillante G (2013) Identification and evaluation of the key social competencies for Chinese construction project managers. Int J Proj Manag 31(5):748–759

AI-Enabled Multi-Project Governance: A Proposed Framework

Erik Araya Aliaga(✉), Costanza Mariani, and Mauro Mancini

Politecnico Di Milano, Milano, Italy
Erikhumberto.araya@polimi.it

Abstract. Amidst rapid technological advancements, AI's application in project management remains underexplored, particularly at the governance level. This study builds on the Task-Technology Fit theory to construct a governance-by-AI framework that aligns AI capabilities with key project governance functions. The proposed framework explains the AI potential for enhancing decision-making, risk management, and stakeholder alignment, improving the effectiveness of project outputs and outcomes. By integrating empirical examples and theoretical insights, this paper contributes to academic literature by providing a framework of how AI allow to detect, predict and simulate in portfolio and project management. We also offer practical guidance for organizations aiming to harness AI's potential. The findings suggest that strategic implementation of AI can lead to substantial improvements in managing complex projects, providing a roadmap for future research and practical applications in the field.

Keywords: Artificial Intelligence · Governance · Multiple Projects · Machine Learning

1 Introduction

Artificial Intelligence refers to the branch of computer science that focuses on creating intelligent machines capable of performing tasks that typically require human intelligence (OECD 2022). In recent years, there have been remarkable advancements in this field, leading to major technological transformations in different fields such as healthcare, finance, transportation, and entertainment (OECD 2019).

A recent survey by the McKinsey Global Institute (McKinsey Global Institute 2022) has shown that in these industries, at least 5% of companies' earnings before interest and taxes (EBIT) was attributable to AI in 2021—highlighting that the investments made in implementing these new technologies pay off in the short term. On the same line, a (PwC 2017) report forecasts that the AI addition to the global economy in 2030 could be more significant than the current output of China and India merged, and an Accenture (2018) report designates that in 2035, AI could double annual growth rates of gross value added in 12 developed countries that currently make up the 50% of global gross domestic product.

E. Favari et al. (Eds.): MeRIT 2024, LNCE 600, pp. 35–46, 2025.
https://doi.org/10.1007/978-3-031-92797-3_3

In this world moving towards a progressive technological transformation, this shift toward artificial intelligence will also likely impact the project management domain. While some practitioner-based studies report that project managers perceive the high potential of AI in project management (Association for Project Management 2022; IPMA and PwC 2020), the academic literature on the adoption of Artificial Intelligence in project management is not very broad and does not present significant results with respect to empirical applications. At present, authors' opinion ranges from some advocating a gradual application of AI in project management (Bodea et al. 2020; Holzmann et al. 2022), to the prediction of a more radical impact over the next few years, especially for the management of complex and challenging projects (Association for Project Management 2022; Fridgeirsson et al. 2021).

Authors who argue for the radical impact of artificial intelligence on project management often refer to the individual project level where the literature reports demonstrations and cases of AI applications. For example, the risk management processes are widely covered: according to (Pan and Zhang 2021), various AI methods, such as probabilistic models, fuzzy theory, machine learning, and neural networks are employed to replace traditional risk analysis methods, (see also Mancini et al. 2023; Mariani and Mancini 2023; Taroun and Yang 2013; Wang and Jin 2019); at the same time also project scheduling (Faghihi et al. 2015), effort estimation (Bisi and Goyal 2016; Elish et al. 2013; Moradbeiky 2023) and cost budgeting (Afzal et al. 2019; Cheng et al. 2015; Kamoona and Budayan 2019) seem to be fertile ground for the application of AI.

Most authors reporting applications at the level of individual project planning and control, outline that activities based on quantitative data, like project cost, scope, and risk management are those that have the potential to benefit the most from AI applications (Fridgeirsson et al. 2021; Ruiz et al. 2020).

For strategic activities, such as those overseen by project governance, focused on managing the outputs, outcomes, and benefits of projects, programs, and portfolios (Müller 2019; Müller et al. 2017; Turner 2006, 2020), are claimed by some authors to be also benefited by AI. Key areas of project governance like monitoring the achievement of project goals, stakeholder satisfaction (Cellerino and Mancini 2023), and alignment with business objectives can be beneficiary of this new advancement (Bodea et al. 2020; Holzmann et al. 2022), enabling more data-driven decision-making (El Khatib and Al Falasi 2010).

This paper aims to identify the potential of applying AI to project governance, proposing a governance-by-AI conceptual framework. To this end, the authors first build a structure of the project governance model on multiple project context based on definitions and grounding on the Task-Technology fit theory (Goodhue and Thompson 1995); second, the paper proposes a conceptual framework positing a positive impact of the application of AI on the fundamental project governance functions.

2 Background

This section compiles the literature essential for defining the framework of multi-level project governance by AI. Specifically, it articulates the concept of governance, outlines the organizational structures supporting enterprise project governance, and highlights the functions that AI can perform in this context.

2.1 Project Governance

Although the literature presents a broad range of definitions, the most widely recognized description of corporate governance's objective is provided by the OECD (2015). According to this definition, the primary purpose of corporate governance is to foster an environment characterized by trust, transparency, and accountability. An important distinction in governance is highlighted in an extensive literature review by (Ahola et al. 2014), in which the authors mention that there are (i) papers that define project governance as internal to a single specific project and (ii) those that describe project governance as external to any specific project. The former (i) views the project, as opposed to its organization, as a powerful organizational actor, which involves many different organizations and can "set the rules of the game". This perspective aligns with the theory of the project as a temporary organization and sees its governance aligned with and shaped by the past and present experiences of the different organizations involved in the temporary endeavor (Engwall and Jerbrant 2003; Lundin and Söderholm 1995). On the other hand, the latter (ii) considers individual projects as subjected to governance by their public or private owners. In this view, project governance is seen as the means to align the goals of single projects conducted simultaneously with the short and long-term strategic goals of the parent organization, ensuring that the project can serve as a vehicle that executes those strategic goals and generate value for the organization (Klakegg 2009; Turner and Keegan 2001).

In the context of this book chapter, in which we conceptualize a Multi-level Governance by AI framework, the perspective of analysis availed is the latter, as the goal is to understand how these innovative technologies can support "the structures, principles, and processes for undertaking the management of projects" (Crawford and Cooke-Davies 2009) – to emphasize the importance, from an "enterprise project management" perspective (Oun et al. 2016), of the synergies attainable in a multi-projects context.

In a multi-project context, project governance has two fundamental functions. The first involves deciding which projects the organization should approve, invest in, and support (Blomquist and Muller 2006; Martinsuo 2013; Teller and Kock 2013), while the second consists of the oversight and assurance that the projects undertaken are effectively managed and that their performance remains aligned with the strategic plan (Martinsuo and Lehtonen 2007).

2.2 Machine Learning in Governance

Machine learning models are a subtype of Artificial Intelligence that are specifically trained to unveil patterns and correlations in large datasets providing support for prediction and decision-making. Depending on the nature of the data and the desired outcome, machine learning can be categorized as supervised or unsupervised and it can be used for predictions, classifications, and data clustering.

In the field of project management literature, AI applications predominantly pertain to quantitative areas of management such as risk management (Islam et al. 2021; Mancini et al. 2023; Mariani and Mancini 2023), progress monitoring (Afzal et al. 2020; Flyvbjerg et al. 2022), and project cost estimation and control (Cheng and Hoang 2014; Cheng et al. 2010a).

However, the firm-level literature also highlights interesting applications from the governance perspective, not limited just to management activities. In particular, a paper by (Margetts 2022) suggests that there are already governmental tasks for which AI has proved helpful. The authors identify detection, prediction, and simulation as general tasks, revealing and reinforcing key governance requirements such as fairness, transparency, and accountability.

Multiple authors (Hanisch et al. 2023; Tassabehji et al. 2016) point out that governance organizational structures need to incorporate the utilization of AI models for detection, defined as tools that gather information from big data to facilitate more informed decision-making. AI-powered advancements in robotics, computer vision, and spatial computing can enhance detection, aiding for example in the early identification of fraud or financial scams (Margetts 2022).

Machine learning has much to offer also for prediction applications to governance structures, serving as a valuable tool for foresight and forecasting. In a general sense, the utilization of predictive risk-based models can improve the monitoring of project progress, enabling regular inspection and oversight updates. Additionally, AI tools can forecast aggregate demand, aiding in the selection of projects for the portfolio (Papagiannidis et al. 2022).

Finally, governance organizational bodies can leverage the *simulation* capabilities of AI to design interventions and regulations that are driven by evidence and rich in data. Simulations can support conducting tests on interventions before implementation to comprehend their probable effects, particularly those related to expensive new initiatives, significant changes in resource allocation, or cost-cutting measures in programs and projects (Margetts 2022).

While we are aware that there may be additional actions that AI can support, in the scope of this paper, we will limit the analysis to highlighting the impact that detection, prediction, and simulation can have. In addition to specifically choosing these three actions because they are highlighted as predominant functions of AI in abundant existing literature (Guenduez and Mettler 2023; Haenlein and Kaplan 2019; Howard 2019), we decided to incorporate them as they best represent the potential functions of AI specifically in governance (Fenwick et al. 2019; Hilb 2020).

3 A Governance Framework

Building on the definitions of governance provided by (Blomquist and Muller 2006; Martinsuo 2013; Martinsuo and Lehtonen 2007; Teller and Kock 2013), for building our framework we consider two main governance functions: Portfolio Management (Blomquist and Muller 2006; Martinsuo 2013; Teller and Kock 2013), where the organization needs to decide the projects to be included in the portfolio, and the resource distribution across them. The second function is the governance and oversight of ongoing projects (Project Management) (Martinsuo and Lehtonen 2007). The primary objective of this function is to oversee projects to ensure their performance aligns with the plan and to prevent the overuse of resources that could otherwise be allocated to other projects in the portfolio.

Figure 1, shows the interrelationship among different governance functions. Project control at the individual project level is directly influenced by resource allocation decisions made at the portfolio level. Simultaneously, resource management within each project depends on performance outcomes, which are shaped by the effectiveness of project control measures. If a project underperforms, resource distribution at the portfolio level may be adjusted to better support it, potentially influencing the selection of future projects as well.

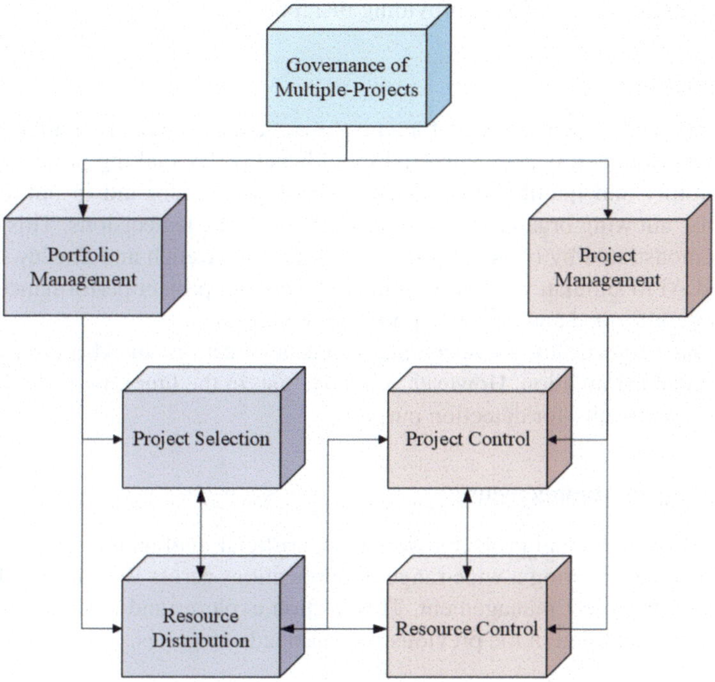

Fig. 1. Governance in Multiple-Projects context

4 Uses of AI in Governance

4.1 AI in Portfolio Management

In this framework, the focus is on the applications and optimization of AI-driven processes rather than specifying particular AI techniques. Given the rapid advancements in AI over recent years, emphasizing specific types of AI could limit generalization and reduce the framework's relevance over time. Therefore, drawing from (Margetts 2022), we classify AI within this framework into three primary categories: detection, prediction and simulation.

4.1.1 Prediction

In the context of prediction, the literature outlines its applications across various phases, notably during the bid-no bid phase, where it assists in forecasting the likelihood of a project bid's success by analyzing historical data. For example, (Sonmez and Sözgen 2017) demonstrated how AI could predict bid/no bid decisions for Oil & Gas projects by assessing factors such as project scope alignment, personal relationships, and price levels. Additionally, AI can enhance the accuracy of cost estimates crucial for project bidding. Cheng et al. (2010b) proposed an evolutionary fuzzy hybrid neural network to predict conceptual cost estimates, providing insights.

4.1.2 Simulation

In the evaluation of project selection for a portfolio, AI-based models simulate potential project outcomes to assess associated risks and benefits. By ranking projects based on their profitability and likelihood of success, these simulations aid in optimizing the portfolio mix, allowing organizations to make more strategic decisions. This approach has been demonstrated by (Costantino et al. 2015), and (Relich and Pawlewski 2017), who utilized AI to simulate success rankings and predict project performance, thereby enabling more informed and effective portfolio management.

In the case of portfolio management, current applications of AI are only focused on prediction and simulation. However, a notable gap in the literature is the lack of AI applications specifically for detection purpose.

4.2 AI in Project Management

In the context of individual project governance, artificial intelligence (AI) has become an invaluable asset, offering a wide range of applications across the various phases and tasks involved in project management. This section explores and categorizes these AI applications, as identified in the previously mentioned categories.

4.2.1 Detection

In the domain of Risk Assessment, AI tools such as Bayesian Belief Networks and Fuzzy Hybrid Methods are instrumental in detecting potential risks by analyzing project data and identifying hidden patterns that may otherwise go unnoticed. These methods facilitate a more proactive approach to risk management by leveraging probabilistic reasoning and fuzzy logic to handle uncertainty, thus enabling project teams to recognize and address potential threats early in the project lifecycle (Khakzad et al. 2013; Li et al. 2013).

Additionally, Structural Equation Modeling (SEM) is widely used within Risk Assessment to not only identify relationships among various risk factors but also to predict their potential impacts on project outcomes. By employing SEM, project managers can gain a deeper understanding of complex risk interdependencies and how they influence one another. This advanced modeling approach allows for the exploration of causal relationships, enhancing the ability to forecast how certain risks may propagate through the project and ultimately affect its success (Yildiz et al. 2014).

4.2.2 Prediction

In the realm of schedule and duration prediction, Artificial Neural Networks (ANN) and Support Vector Machines (SVM) have demonstrated considerable promise in forecasting project timelines from the initial planning stages. For example, ANN models have been used to forecast project durations by analyzing complex, nonlinear variables (Wang et al. 2012). Meanwhile, SVMs excel in identifying hidden patterns within planning data that can influence project duration and scheduling (Wauters and Vanhoucke 2014). With these models, project managers can visualize potential project timelines at an early stage, which allows for better-informed decisions regarding resource allocation and project pacing.

Cash flow forecasting has also benefited from AI advancements. Models like the Adaptive Time-dependent Least Squares Support Vector Machine allow for dynamic cash flow projections throughout a project's lifecycle. This model, as discussed by (Cheng et al. 2015), adjusts to project-specific factors over time, delivering more accurate, adaptive cash flow forecasts. In the same economic context, AI-driven cost estimation tools, particularly those using Artificial Neural Networks, provide a robust framework for forecasting expenditures across different project phases. For instance, (Barakchi et al. 2017) applied ANN models to cost estimation, allowing for accurate predictions that consider complex interdependencies, such as labor and material costs, as well as historical spending trends.

In terms of outcome prediction, algorithms like unsupervised random forests and classification models enable project managers to gauge current project performance against historical benchmarks (Flyvbjerg et al. 2022).

Predictive techniques combining Genetic Algorithms with Deep Neural Networks and SVM algorithms have been employed to forecast the Estimate at Completion (EAC). These hybrid models, as explored by (Cheng et al. 2010b; Kamoona and Budayan 2019), aid in identifying potential risks and delays by modeling scenarios that impact final project outcomes.

Accurate forecasting of resource requirements, particularly concerning labor and materials, is vital for ensuring project efficiency. ANN models (Bisi and Goyal 2016) and other machine learning algorithms, like those applied by (Polkowski et al. 2019), can predict resource needs based on specific project tasks. These models analyze factors such as task complexity, duration, and required skill levels to forecast necessary resources accurately.

4.2.3 Simulation

Optimization algorithms such as Genetic Algorithms (GA) and Ant Colony Optimization (ACO) enhance resource allocation by simulating potential project scenarios within constraints like budget and time. GA applies natural selection principles to refine distribution plans iteratively, while ACO models ant behavior to explore project paths, identifying the most efficient routes (Cheng and Yan 2009; Peng and Liu 2022).

For risk response, such algorithms model various scenarios to determine the best strategies under budget, time, and risk constraints, helping minimize delays and cost overruns (Shoar and Nazari 2019; Zabel et al. 2014).

In contingency budgeting, Fuzzy Analytic Hierarchy Process (Fuzzy-AHP) and Artificial Neural Networks (ANN) estimate contingency funds under uncertain conditions, improving risk evaluation accuracy and ensuring resource availability for unexpected challenges (Afzal et al. 2020).

5 AI and Multi-project Governance Framework

In Fig. 2, the proposed framework illustrates the various applications of AI designed to enhance the processes involved in multi-project governance. At the top level, the framework begins with the overarching theme of governance of multiple projects, emphasizing the need for structured oversight in managing several projects simultaneously. The framework branches into two main components: Portfolio Management and Project Management.

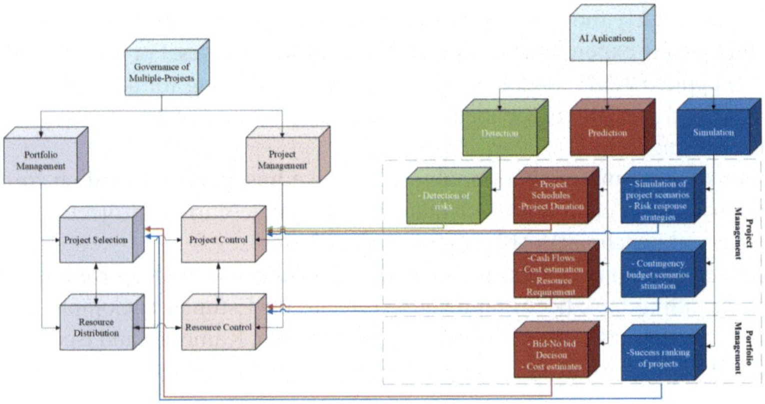

Fig. 2. AI in Multi-project Governance Framework

The Portfolio Management segment includes critical functions such as Project Selection and Resource Distribution, both interdependent as the selection of projects is done based on the resources available and the distribution of resources and vice-versa.

Within project management, project control significantly influences the performance of individual projects and the management of allocated resources. As resources are utilized, project control must adapt accordingly. Additionally, unforeseen circumstances may necessitate further resource allocation, thereby impacting the broader project portfolio. As demonstrated in the framework, AI can significantly enhance decision-making, predictive capabilities, and standardization across various aspects of governance. For example, AI applications can improve decision-making by analyzing historical data to forecast future trends and enabling real-time adjustments to project execution. Additionally, predictive analytics can be utilized to forecast project duration and budget needs, ensuring that managers can proactively address potential challenges.

The inclusion of risk management strategies highlights the importance of AI in identifying and mitigating risks associated with project execution. This allows project managers to implement contingency plans that are informed by AI-driven insights, thereby enhancing overall project resilience.

The framework also addresses resource control, by employing AI techniques such as optimization algorithms and machine learning models. In this way organizations can better understand their resource needs, ensuring optimal utilization and minimizing waste.

References

1. Accenture (2018) Artificial Intelligence genuine impact
2. Afzal F, Yunfei S, Junaid D, Hanif MS (2020) Cost-risk contingency framework for managing cost overrun in metropolitan projects: using fuzzy-AHP and simulation. Int J Manag Proj Bus 13(5):1121–1139. https://doi.org/10.1108/IJMPB-07-2019-0175
3. Afzal F, Yunfei S, Nazir M, Bhatti SM (2019) A review of artificial intelligence based risk assessment methods for capturing complexity-risk interdependencies: cost overrun in construction projects. Int J Manag Proj Bus. https://doi.org/10.1108/IJMPB-02-2019-0047
4. Ahola T, Ruuska I, Artto K, Kujala J (2014) What is project governance and what are its origins? Int J Proj Manag 32(8):1321–1332. https://doi.org/10.1016/j.ijproman.2013.09.005
5. Association for Project Management (2022) Artificial intelligence in project management
6. Barakchi M, Torp O, Belay AM (2017) Cost estimation methods for transport infrastructure: a systematic literature review. Proc Eng 196:270–277. https://doi.org/10.1016/j.proeng.2017.07.199
7. Bisi M, Goyal NK (2016) Software development efforts prediction using artificial neural network. IET Softw 10(3):63–71. https://doi.org/10.1049/iet-sen.2015.0061
8. Blomquist T, Muller R (2006) Practices, roles, and responsibilities of middle managers in program and portfolio management. Proj Manag J 37(1):52–66
9. Bodea C, Mitea C, Stanciu O (2020) Artificial Intelligence adoption in project management: main drivers, barriers and estimated impact. In: Proceedings of the 3rd international conference on economics and social sciences, pp 758–767. https://doi.org/10.2478/9788366675162-075
10. Cellerino F, Mancini M (2023) Stakeholder involvement in the development of megaprojects: a dedicated framework for transport infrastructure. In: Favari E, Cantoni F (eds) Complexity and Sustainability in Megaprojects. MERIT 2022, Lecture Notes in Civil Engineering, vol 342. Springer, Cham, pp 77–95. https://doi.org/10.1007/978-3-031-30879-6_7
11. Cheng MY, Hoang ND (2014) Interval estimation of construction cost at completion using least squares support vector machine. J Civ Eng Manag 20(2):223–236. https://doi.org/10.3846/13923730.2013.801891
12. Cheng MY, Tsai HC, Sudjono E (2010a) Conceptual cost estimates using evolutionary fuzzy hybrid neural network for projects in construction industry. Expert Syst Appl 37(6):4224–4231. https://doi.org/10.1016/j.eswa.2009.11.080
13. Cheng M-Y, Peng H-S, Wu Y-W, Chen T-L (2010b) Estimate at completion for construction projects using evolutionary support vector machine inference model. Autom Constr 19(5):619–629. https://doi.org/10.1016/j.autcon.2010.02.008
14. Cheng M-Y, Hoang N-D, Wu Y-W (2015) Cash flow prediction for construction project using a novel adaptive time- dependent least squares support vector machine inference model. J Civ Eng Manag 21(6):679–688. https://doi.org/10.3846/13923730.2014.893906

15. Cheng T, Yan R (2009) Integrating messy genetic algorithms and simulation to optimize resource utilization. Comput Aided Civ Inf Eng 24:401–415. https://doi.org/10.1111/j.1467-8667.2008.00588.x
16. Costantino F, Di Gravio G, Nonino F (2015) Project selection in project portfolio management: an artificial neural network model based on critical success factors. Int J Proj Manag 33(8):1744–1754. https://doi.org/10.1016/j.ijproman.2015.07.003
17. Crawford L, Cooke-Davies T (2009) Project governance: the role and capabilities of the executive sponsor. Proj Perspect, XXXI, pp 66–74
18. Elish MO, Helmy T, Hussain MI (2013) Empirical study of homogeneous and heterogeneous ensemble models for software development effort estimation. Math Probl Eng 2013:1–21
19. Engwall M, Jerbrant A (2003) The resource allocation syndrome: the prime challenge of multi-project management? Int J Proj Manag 21(6):403–409. https://doi.org/10.1016/S0263-7863(02)00113-8
20. Faghihi V, Nejat A, Reinschmidt KF, Kang JH (2015) Automation in construction scheduling: a review of the literature. Int J Adv Manuf Technol 81(9–12):1845–1856. https://doi.org/10.1007/s00170-015-7339-0
21. Fenwick M, McCahery JA, Vermeulen EPM (2019) The end of 'corporate' governance: hello 'platform' governance. Eur Bus Organ Law Rev 20(1):171–199. https://doi.org/10.1007/s40804-019-00137-z
22. Flyvbjerg B et al. (2022). AI in action: how the Hong Kong development bureau built the pss — an early-warning-sign system for public works projects. SSRN, 1–28
23. Fridgeirsson TV, Ingason HT, Jonasson HI, Jonsdottir H (2021) An authoritative study on the near future effect of artificial intelligence on project management knowledge areas. Sustainability (Switzerland) 13(4):1–20. https://doi.org/10.3390/su13042345
24. Goodhue DL, Thompson RL (1995) Task-technology fit and individual performance. MIS Q Manag Inform Syst 19(2):213–233. https://doi.org/10.2307/249689
25. Guenduez AA, Mettler T (2023) Strategically constructed narratives on artificial intelligence: what stories are told in governmental artificial intelligence policies? Gov Inf Q 40(1):101719. https://doi.org/10.1016/j.giq.2022.101719
26. Haenlein M, Kaplan A (2019) A brief history of artificial intelligence: on the past, present, and future of artificial intelligence. Calif Manag Rev 61(4):5–14. https://doi.org/10.1177/0008125619864925
27. Hanisch M, Goldsby CM, Fabian NE, Oehmichen J (2023) Digital governance: a conceptual framework and research agenda. J Bus Res 162(February):113777. https://doi.org/10.1016/j.jbusres.2023.113777
28. Hilb M (2020) Toward artificial governance? The role of artificial intelligence in shaping the future of corporate governance. J Manag Gov 24(4):851–870. https://doi.org/10.1007/s10997-020-09519-9
29. Holzmann V, Zitter D, Peshkess S (2022) The expectations of project managers from Artificial Intelligence: a Delphi study. Proj Manag J 53(5):438–455. https://doi.org/10.1177/87569728211061779
30. Howard J (2019) Artificial intelligence: implications for the future of work. Am J Ind Med 62(11):917–926. https://doi.org/10.1002/ajim.23037
31. IPMA, PwC (2020) Artificial intelligence impact in project management
32. Islam MS, Nepal MP, Skitmore M, Drogemuller R (2021) Risk induced contingency cost modeling for power plant projects. Autom Constr 123:103519. https://doi.org/10.1016/j.autcon.2020.103519
33. Kamoona K, Budayan C (2019) Implementation of genetic algorithm integrated with the deep neural network for estimating at completion simulation. Adv Civil Eng 2019(ID 7081073):1–15

34. Khakzad N, Khan F, Amyotte P (2013) Quantitative risk analysis of offshore drilling operations: a Bayesian approach. Saf Sci 57:108–117. https://doi.org/10.1016/j.ssci.2013.01.022
35. El Khatib M, Al Falasi A (2010) Effects of artificial intelligence on decision making in project management. Am J Indus Busin Manag 11(03):251–260. https://doi.org/10.4236/ajibm.2021.113016
36. Klakegg OJ (2009) Pursuing relevance and sustainability improvement strategies for major public projects. Int J Manag Proj Bus 2(4):499–518. https://doi.org/10.1108/17538370910991115
37. Li F, Phoon KK, Du X, Zhang M (2013) Improved AHP method and its application in risk identification. J Constr Eng Manag 139(3):312–320. https://doi.org/10.1061/(asce)co.1943-7862.0000605
38. Lundin RA, Söderholm A (1995) A theory of the temporary organization. Scand J Manag 11(4):437–455. https://doi.org/10.1016/0956-5221(95)00036-U
39. Mancini M, Mariani C, Manfredi M (2023) Nuclear decommissioning risk management adopting a comprehensive artificial intelligence framework: an applied case in an Italian site. Prog Nucl Energy 158(February):104589. https://doi.org/10.1016/j.pnucene.2023.104589
40. Margetts H (2022) Rethinking AI for good governance. Daedalus 151(2):360–371. https://doi.org/10.1162/DAED_a_01922
41. Mariani C, Mancini M (2023) Selection of projects' primary and secondary mitigation actions through optimization methods in nuclear decommissioning projects. Nucl Eng Des 407(March):112284. https://doi.org/10.1016/j.nucengdes.2023.112284
42. Martinsuo M (2013) Project portfolio management in practice and in context. Int J Proj Manag 31(6):794–803. https://doi.org/10.1016/J.IJPROMAN.2012.10.013
43. Martinsuo M, Lehtonen P (2007) Role of single-project management in achieving portfolio management efficiency. Int J Proj Manag 25(1):56–65. https://doi.org/10.1016/J.IJPROMAN.2006.04.002
44. McKinsey Global Institute (2022) The state of AI in 2022—and a half decade in review
45. Moradbeiky A (2023) FEEM: a flexible model based on artificial intelligence for software effort estimation. J AI Data Min 11(1):39–51. https://doi.org/10.22044/jadm.2022.11161.2265
46. Müller R (2019) Governance, governmentality and project performance: the role of sovereignty. Int J Inf Syst Proj Manag 7(2):5–17
47. Müller R, Zhai L, Wang A (2017) Governance and governmentality in projects: profiles and relationships with success. Int J Proj Manag 35(3):378–392. https://doi.org/10.1016/j.ijproman.2017.01.007
48. OECD (2015) G20/OECD principles of corporate governance. OECD
49. OECD (2019) Artificial intelligence in society. OECD
50. OECD (2022) Harnessing the power of AI and emerging technologies (340)
51. Oun TA, Blackburn TD, Olson BA, Blessner P, Oun TA, Blackburn TD, Olson BA, Blessner P, Blackburn TD, George T, Olson BA, George T, Blessner P, George T (2016) An enterprise-wide knowledge management approach to project management. Eng Manag J 28(3):179–192. https://doi.org/10.1080/10429247.2016.1203715
52. Papagiannidis E, Enholm IM, Dremel C, Mikalef P, Krogstie J (2022) Toward AI governance: identifying best practices and potential barriers and outcomes. Inf Syst Front:123–141. https://doi.org/10.1007/s10796-022-10251-y
53. Peng J, Liu X (2022) Labor resource allocation under extremely short construction period based on the inverse optimization method. Eng Constr Archit Manag 1:1254–1271. https://doi.org/10.1108/ECAM-06-2022-0604
54. Polkowski Z et al. (2019) Machine learning-based software effort estimation: an analysis, 1–6

55. PwC (2017) Sizing the prize What's the real value of AI for your business and how can you capitalise? (9)
56. Ruiz JG, Torres JM, Crespo RG (2020) The application of artificial intelligence in Project Management research: a review. Int J Interact Multimedia Artif Intell 6(6):54–66. https://doi.org/10.9781/ijimai.2020.12.003
57. Shoar S, Nazari A (2019) An optimization framework for risk response actions selection using hybrid ACO and FTOPSIS. Sci Iran 26(3E):1763–1777. https://doi.org/10.24200/sci.2018.20225
58. Sonmez R, Sözgen B (2017) A support vector machine method for bid/no bid decision making. J Civ Eng Manag 23(5):641–649. https://doi.org/10.3846/13923730.2017.1281836
59. Taroun A, Yang J-B (2013) A DST-based approach for construction project risk analysis. J Oper Res Soc 64(8):1221–1230. https://doi.org/10.1057/jors.2013.38
60. Tassabehji R, Hackney R, Popovič A (2016) Emergent digital era governance: enacting the role of the 'institutional entrepreneur' in transformational change. Gov Inf Q 33(2):223–236. https://doi.org/10.1016/j.giq.2016.04.003
61. Teller J, Kock A (2013) An empirical investigation on how portfolio risk management influences project portfolio success. Int J Proj Manag 31(6):817–829. https://doi.org/10.1016/J.IJPROMAN.2012.11.012
62. Turner JR, Keegan A (2001) Mechanisms of governance in the project-based organization: roles of the broker and steward. In. Eur Manag J 19(3):254–267
63. Turner R (2006) Towards a theory of project management: the nature of the project governance and project management. Int J Proj Manag 24:93–95. https://doi.org/10.1016/j.ijproman.2005.11.008
64. Turner R (2020) How does governance influence decision making on projects and in project-based organizations? Proj Manag J 51(6)
65. Wang Y, Jin X (2019) Structural risk of diversified project financing of city investment company in China based on the best worst method. Eng Construct Architect Manag:51608363. https://doi.org/10.1108/ECAM-05-2019-0249
66. Wang Y-R, Yu C-Y, Chan H-H (2012) Predicting construction cost and schedule success using artificial neural networks ensemble and support vector machines classification models. Int J Proj Manag 30(4):470–478. https://doi.org/10.1016/j.ijproman.2011.09.002
67. Wauters M, Vanhoucke M (2014) Support Vector Machine Regression for project control forecasting. Autom Constr 47:92–106. https://doi.org/10.1016/j.autcon.2014.07.014
68. Yildiz AE, Dikmen I, Birgonul MT, Ercoskun K, Alten S (2014) A knowledge-based risk mapping tool for cost estimation of international construction projects. Autom Constr 43:144–155. https://doi.org/10.1016/j.autcon.2014.03.010
69. Zabel NY, Georgy ME, Ibrahim ME (2014) Employing ant colony for the optimal reduction of project risk severity. In: 31st international symposium on automation and robotics in construction and mining, ISARC 2014 - proceedings, pp 398–403. https://doi.org/10.22260/isarc2014/0053

Change of Expectations of the Socio-economic Context in the Evaluation of Mega Projects. Risks and Opportunities of Eclecticism Facing Unrealistic Approaches

Massimo Bianchi(✉)

University of Bologna, 40125 Bologna, BO, Italy
massimo.bianchi@unibo.it

Abstract. The situation in progress of megaprojects must consider many diffused changes in the socio-economic context, most of them not favorable to a successful conclusion of works. The more extensive changing is regarding the geo political dimension and the increasing of the role, in GDP and Research, of BRICS and Arab Countries that, by the way, are interested by a growing rate of Mega Projects in progress, submission or only as proposals.

In this challenging context it is increasing the variety of answers from project management conducting to a widespread eclecticism with different and sometimes less coherent tools of intervention.

This led, according to a model of analysis yet applied in past century on politics and economics by Antonio Gramsci and Federico Caffè with their critics, to eclecticism and unrealistic or wishful decisions.

As the difficulties of project management in last decades the paper examines if this model can be applied also to Project Management and if it is fruitful to reduce, in the evaluation of mega projects, unrealistic and inappropriate decision making.

Keywords: Megaprojects · project management · eclecticism · unrealistic decisions · unwritten rules

1 Mega Projects and the Context

It has been considered that the socioeconomic project environment has a relevant relationship with megaproject performance [1].

The situation in progress of megaprojects must consider many diffused changes in the socio-economic context, most of them not favorable to a successful conclusion of works. Increasing difficulties are coming from many different origins: contradictory interpretations about the effects of global warming, the uncertainties in the policies of decarbonization, the surge of doubts on the way to apply green deal, and others [2]. Furthermore, in the macro dimension, it is imposing the reduction of relevance of western countries comparing to BRICS and Arab Countries as it concerns GPD, the manufacturing production [3] and the research.

E. Favari et al. (Eds.): MeRIT 2024, LNCE 600, pp. 47–54, 2025.
https://doi.org/10.1007/978-3-031-92797-3_4

In the last decades most of mega projects was located externally to western countries. Almost all of these projects, received a diffused and strong hostility that seems to be, as it concerns Italy, the bridge on the Messina Strait, the location of gasifiers and the waste-to-energy plants, just to note the main.

This is a general problem but, within it, the limits of the project management must be considered when, instead of resorting to its own methods, it tends to extend its competencies to psychology, sociology, information technology, even neuroscience, to giving a response to which it ultimately turns out to be incompetent.

In truth, on the one hand, management was born as a multidisciplinary subject in order to solve most of the variety of problems that happen within an organization. At the same time, in the evolution of managerial practices, the Peter Principle was formulated, as the achievement of the maximum level of incompetence in the process of managerial development within organizations [4].

2 Eclecticism and Unrealistic Hypothesis

Eclecticism can be related mainly to two different analytical tools concerning the evolution of economic and managerial theories. In last times Eclecticism was studied by Dunning as the eclectic paradigm, also known OLI (Ownership, Location, Internationalization) to analyze the companies' behavior in the pursuing of foreign direct investment [5].

In our analysis the concept of eclecticism is used to analyze the contemporary managerial evolution, coupled with the one of wishful thinking, as once applied by Gramsci and Caffè in the critics of the political-economic thinking.

The basic hypothesis of this analysis is that, in the management, the contemporary flourishing of contributes from different and heterogeneous disciplines, to face applied problems and although the basic multidisciplinary approach of the management, was not completely controlled particularly by practitioners and led to wishful thinking and unrealistic decisions.

To this regard Peter Principle states that, within organizations, each person progresses up the hierarchy and faces increasingly complex problems until he reaches his maximum level of incompetence for which he cannot make further advancements. At this point it can also be said that, as the career progresses, the decisions that the person should make are progressively based on inadequate or unrealistic hypotheses or interpretations.

Some scholars, coming from other disciplinary fields, have formulated a different interpretation of incompetence. They connect it to the evolution of the theories of a specific scientific area which, in certain situations, progresses towards eclecticism, introducing methodologies or approaches specific to other disciplines over which there is no control. Indeed, precisely because of their extemporaneousness, it may happen that these eclectic approaches obtain greater consensus from both a theoretical and practical point of view.

The extension of Peter's Principle to the management of megaprojects concerns the progressive incompetence that occurs when the size of the projects in which one is involved, increases and exceeds the skills and abilities of the manager, imposing a complexity of decisions that he or she is unable to adequately control and decide upon.

The inadequacy of theories and practices in economics is the case of the model of analysis expressed in 20″ years of previous century by Antonio Gramsci and in 50″ years by the Economist Federico Caffè which use the term "velleitarismo" [6] to indicate a wishful or unrealistic thinking as the result of the eclecticism diffused in studies of Economics.

The subject was the inappropriate approach of Political and Economic theories to problems, emerging respectively after the first- and second-World War. Both scholars coupled the diffused eclecticism of Economics with the unrealistic proposals produced to face Italians structural problems, till now unresolved, like the underdevelopment of the south of the country and the well-known "questione meridionale". Furthermore, Caffè extended this analysis to the European situation in which, the scarce applying of Keynesian theories in favor of unrealistic approaches, provoked in 30th years, a dramatic world crisis firstly economic and after political [7].

In recent times, eclecticism, as a category of analysis, has recently been introduced into Economics by Galtung who considers it a key factor for a profound, utopian change in economics and politics [8]. After, Dunning considered eclecticism in a holistic perspective useful for the analysis of the internationalization of companies and their projects in the global context [9] and a paradigm for cross disciplinary analysis [10].

3 Obstacles and Emergencies

In the evolution of the tools applied by management, one cannot avoid asking ourselves what the origin of the eclecticism is brought into play by the trans, cross, inter and poly disciplinary approaches [11] which even transdisciplinary analysis struggles to coordinate in an adequate framework [12].

The same tools proposed to reduce the risk of failure end up increasing complexity by including, in addition to objective parameters, a numerous list of subjective ones [13, 14].

As in theoretical work it has been observed, when this strong criticism or inhibition to proceed derives from doctrinal obstacles and academic opposition, this can lead to the search for original and extemporaneous, in a word eclectic, ways to overcome the difficulties.

The proliferation of theories and hypotheses is part of the decision-making process facing emergencies but, at this point, also unique or exceptional events such as mega projects can be submitted to this phenomenon [15].

To this purpose a tendency till now registered, to deal with the difficulties in maintaining the original budget in front of increasing costs, is the surge of controls, indexes of performance, analytical descriptions and logical matrix, all referred to different fields of disciplinary competence [16].

This is a real challenge also in the submission of projects proposals and mostly is dealt with the introduction of subjects and topics drawn eclectically from various disciplines that, also the trans disciplinary analysis cannot coordinate satisfactorily [17]. This increases the risk of formulation of unrealistic forecasts which can later lead many mega projects to the failure [18].

4 The Context of Mega Project

In recent decades, the incidence of the BRICS and Arab Countries in the presentation of Mega Projects increased [19, 20] and, in parallel, in the last two decades, the BRICS and the Arab countries have grown to become the world's largest producers of scientific knowledge with a remarkable contribution of China (Fig. 1).

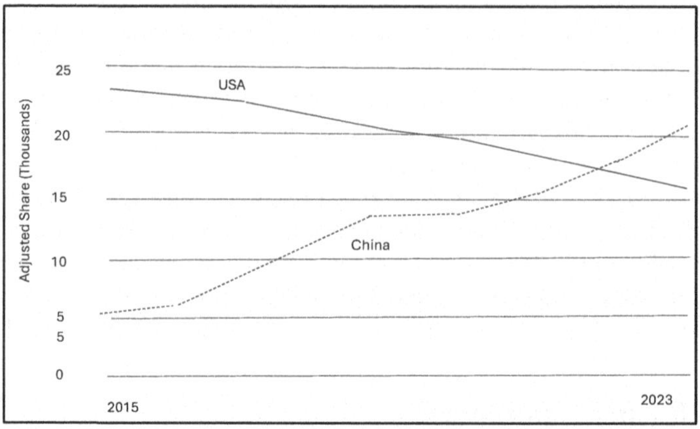

Fig. 1. Trends in Nature index citations 2015–2023 [17]

According to the Digital Science's Dimensions database, last year there were almost 830,000 papers featuring researchers based in China, representing around 15% of the world's 5.4 million articles. These trends show a change in the context in which the management of mega projects operates, a context that manifests itself in an increasingly challenging environment [21].

To this purpose it's significant the minimum and variety of information needed for the preparation of decisions (Fig. 2), a minimum that, coming from different fields of disciplines, produces once more the evidence of relevant eclecticism together with the difficulty to dominate the project management and the increasing risk of unrealistic decisions [22].

This dimension reflects the limits of managerial choices and the current validity of Graiciunas's principle that from one side teaches us the minimum of subordinates that a hierarchical position is in condition to control adequately but from the other side implies a risk of failure in the control of auditors, included the ones involved in mega projects [24].

This complexity, along with the advances of information technology and the current perspectives opened by AI, have led to the hypothesis of models of virtual organizations whose workspace is not necessarily limited to the same area but is made up of a virtual workspace.

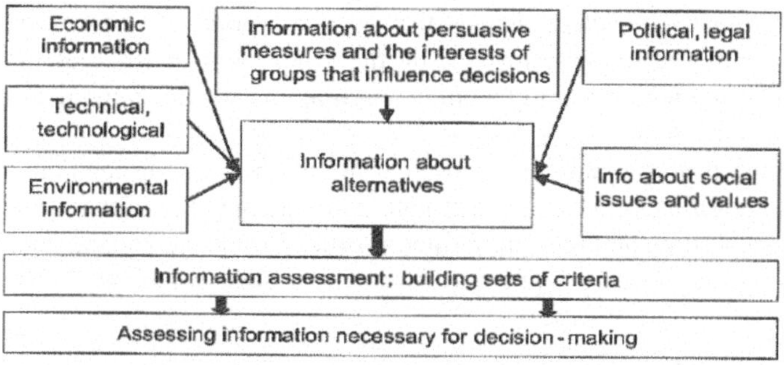

Fig. 2. Variety of information needed for the preparation of decisions [23]

5 The Approach to Mega Projects in Italy

In last times some scholars pointed up on the parallel progress of de-industrialization process and of the crisis of management in Italy with the absence of major projects towards which a mixture of skepticism and mistrust is widespread seems to have been created [25].

From one side, as authoritatively stated, the structure of strategic and advanced sectors of Italian Industry was disjointed and /or badly sold to companies of other countries moving to the international economy the control and the initiative of the national development [26].

At the other side it seems that the Sciences of Management developed in Italy, mainly represented by the Business Management, arc struggling to keep up with the changes induced by the technological and socio-economic revolution.

To be mentioned, the AIDEA Conference of Taormina in which, only in 1982, was officialized the disciplinary location of Managerial Studies [27]. After, in last decades, management researchers must face the uncertainties resulting from the multiplication of inter, trans and poly disciplinary proposals to which the multi-disciplinarity, originally claimed by Business Management doctrine, seems offer no longer an adequate answer.

In this perspective, the eclecticism that emerges from the evolution of studies didn't reach, almost till now, an adequate and shared level of maturity to represent a stimulating and reliable paradigm of reference and also the major criticism on management complexity didn't propose a convincing alternative [28].

6 Is Emerging a New Market for Mega-projects?

The crisis of the Western manufacturing industry, highlighted by the growth of the BRICS and Arab Nations, pushes the emerging of an alternative market to the one regulated by the WTO and characterized by growing constraints in all directions, included the ones which increases the difficulties of the project management.

Sanctions, protection of original productions, duties, ethical-legal constraints, rules of the green deal, recall the Pareto's criticism on protectionism in favor of the free market and the debate on which Federico Caffè gave his passionate contribution [29].

On the one hand, the eclecticism of the solutions proposed in theoretical-practical terms to overcome the impasse thrives, on the other hand, it emerges the unrealistic contradiction between increasingly complex regulations and the deregulation of parallel and unregulated markets.[1]

Scholars and professionals pay particular attention to this unregulated area of the markets and, regarding project management, some authors recommend understanding the gap between written and unwritten rules as crucial for project managers, because, as a well-quoted phrase note from practice: "unwritten rules always win" [30].

7 Written and Unwritten Rules

This could be a very tricky subject but, on the other hand, the widespread gap between original budget and final results of mega projects is a real item and would be unrealistic not consider it as a solid subject [31].

Where the gap between them is greatest, things seem political because we say one thing but reward the other. If you're in a leadership role, see this as a challenge: Where are the gaps between written and unwritten rules in the submission of a mega project and in its realization? What can we do to close these gaps?

Some of them express their empirical nature and once more confirm the need, in project management, of the habit of transforming the project consortium into a team of interacting people, a habit that can be enforced by the didactical approaches like practice enterprise with a prevalence of learning by doing [32].

8 Changing Expectations

As universally stated, complex projects have low success rates. Normally international estimates are in the order of 1/1000 for economic success but extensively practitioners consider "The iron law of mega-projects has become 'over-time, over-budget and over again" [33].

Accordingly, to these data it seems successful for a commitment if the weighted percentage of the different risk factors reaches the 50% and to this purpose a way to build, around the initiative, a community project with the participation of stakeholders and a strong and dynamic relationship with the project team and the top management.

This community must be adapted to the nature of the project. Complicated mega projects with a higher risk profile need a disciplined partner network and a business model adaptable to unknown aspects when they become clear. Once more is relevant to consider the gap we spoke before between written and unwritten rules of behavior and practical management, both complicit in making the path of megaprojects even more complex and the inadequacy of management in dominating them satisfactorily.

[1] See [19]

References

1. Locatelli G, Mikic M, Kovacevic M, Brookes N, Ivanisevic N (2017) The successful delivery of megaprojects: a novel research method. Proj Manag J 48(5):78–94
2. Alcácer J, Cantwell J, Piscitello L (2016) Internationalization in the information age: a new era for places, firms, and international business networks? J Int Bus Stud 47:499–512. https://doi.org/10.1057/jibs.2016.22
3. Bianchi M (2024) Deindustrialization and management sciences in Italy. In: The shadow of velleitarism. 11 Convegno ATINER 6th - 9th may conference for the 11th annual international conference on business, law & economics
4. Peter JL (1969) Il principio di Peter: perché le cose vanno sempre male. William Morrow and Company, ISBN 978-0061699061
5. Sharmiladevi JC (2017) Understanding dunning's oli paradigm. Ind J Commerce Manag Stud VIII(3) ISSN: 2249-0310 EISSN: 2229-5674
6. Poettinger M (2016) Velleitarismo, il giudizio storico di Federico Caffè sul secondo dopoguerra, Ricerche di storia economica e sociale, Vol. II. 1–2, p. 179, https://doi.org/10.17426/55366 ©2017 Università Roma Tre-CROMA
7. Caffè F (2014) La dignità del lavoro. Lit Edizioni, Roma, p 36
8. Galtung J (1967) Theory and methods of social research. Universitetsforlaget, Oslo
9. Dunning JH (1992) Multinational enterprises and the global economy. Addison-Wesley Publishing Company
10. Cantwell J (2015) An introduction to the eclectic paradigm as a meta-framework for the cross-disciplinary analysis of international business. In: Cantwell J (ed) The eclectic paradigm. Palgrave Macmillan, London. https://doi.org/10.1007/978-1-137-54471-1_1
11. Bianchi M (2023) Eclettismo e transdisciplinarità negli studi aziendali italiani. Economia Aziendale Online 14(4)
12. Renn O (2021) Transdisciplinarity: Syntesis toward a modular approach. Futures 130., ISSN 0016-3287. https://doi.org/10.1016/j.futures.2021.102744
13. Shekshnia S, Vyas V (2023) Inside the black box crucial to megaproject success. INSEAD Publ
14. Vyas V, Zweifel TD (2022) Gorilla in the cockpit: breaking the hidden patterns of project failure
15. Virtuani R, Barbareschi B, Cantoni F (2023) Overtaking the traditional leadership style in megaprojects. In: Favari E, Cantoni F (eds) Complexity and sustainability in mega projects, Lecture Notes in Civil Engineering, vol 342. Springer., ISSN 2366-2557 ISSN 2366-2565 (electronic) ISBN 978-3-031-30878-9 ISBN 978-3-031-30879-6 (eBook). https://doi.org/10.1007/978-3-031-30879
16. Bianchi M (2022) Practice enterprise and MOOCs in the design and implementation of megaprojects. Some lessons from European Projects. In: Favari E, Cantoni F (eds) Complexity and sustainability in mega projects. Springer
17. Gibbs P, Beavis A (2020) Thematics reflections. In: Gibbs P, Beavis A (eds) Contemporary thinking on transdisciplinary knowledge, Springer Briefs in Education. Springer. https://doi.org/10.1007/978-3-030-39785-2_1
18. Mišić S, Radujkovic M (2015) Critical drivers of megaprojects success and failure. Proc Eng 122:71–80. https://doi.org/10.1016/j.proeng.2015.10.009
19. Ngwenya, M., Sfide socioeomiche nei progetti di mega infrastrutture., WEGlobal, (2023)
20. Patrick S (2024) BRICS expansion, the G20 and the future of work order. Ottobre, Carnegie Endowment for International Peace
21. Nature Indexes (2024) L'influenza della ricerca cinese porta alla crescita della editoria scientifica nazionale. Nature

22. Rahman A, Bridge A, Rowlinson S, Hubbard B, Xia B (2018) Multinational contracting and the eclectic paradigm of internationalization. Eng Construct Architect Manag. ISSN: 0969-9988
23. Sligoriceniene J, Kaklauskas A, Strelmikienè D, Bianchi M (2012, 2012) Multiple criteria decision support system for the assessment of energy generation technologies considering the dimension of values. Int J Strateg Prop Manag 16(4) Routledge, ISSN1648-715
24. Rossignoli C (2004) Coordinamento e cambiamento. Tecnologia e processi interorganizzativi, Franco Angeli, Milano, p 114
25. Bianchi M, Tampieri L (2013) Graiciunas principle and hyperlinks in the prevention of audit failures. A challenge for auditors and ICT audit supports. Int J Auditing Technology 1(3/4):261
26. La Gallino L (2003) scomparsa dell'Italia Industriale. Einaudi, Torino
27. Rebora G, Ferrando PM (2024) Due Manifesti per una riflessione congiunta sulle organizzazioni e sulle imprese in Sicca, L.M., Torre, T. (2024) On the Sidelines of the Manifesto "The Enterprise We Want (A New Enterprise in a New Economy)" Reflections and Provocations for a Debate We Would Like, Impresa Progetto, n. 1, ISSN 1824-3576
28. Giuffrè (1983) AAVV L'organizzazione nella Economia Aziendale, Milano
29. Cavazzuti F (2015) Appunti e spunti da una breve stagione riformista: 1958–1963. Moneta e Credito 68(270):187–218
30. Brownlee D (2019) The unwritten rules of managing-up. Berrett-Koehler Publishers
31. Parker M (2002) Against management. Organization in the age of managerialism. Polity Press, pp 200–209
32. Bianchi M (2022) Practice enterprise and MOOCs in the design and implementation of megaprojects. Some lessons from European Projects. In: Favari E, Cantoni F (eds) Complexity and Sustainability in Mega Projects, Lecture Notes in Civil Engineering, vol 342. Springer., ISSN 2366-2557 ISSN 2366-2565 (electronic) Lecture Notes in Civil Engineering ISBN 978-3-031-30878-9
33. Saxelby D, Dunn M (2019) Changing the game how Australia can achieve success in the new world of mega-projects. Australian Constructor Association

Role of Megaprojects in Achieving Sustainability Transition, and How Do Project Governance and Technological Integration Impacts Megaproject Performance Through Stakeholder Engagement

Muhammad Shoaib Iqbal[1](✉), S. Ping Ho[1], and Nosheen Qadeer[2]

[1] Department of Civil Engineering, National Taiwan University, Da'an District, Taipei 10617, Taiwan
{d11521031,spingho}@ntu.edu.tw

[2] Department of Economics, Faculty of Management and Economics, Tomas Bata University in Zlin, T. G. Masaryka 5555, 76001 Zlín, Czech Republic
qadeer@utb.cz

Abstract. This research examines the relationships between Project Governance, Technological Integration, Stakeholder Engagement, and Project Performance in large-scale construction megaprojects. Drawing on Agency Theory and Stakeholder Theory, the study explores how governance and technology enhance project outcomes, with a focus on the mediating role of stakeholder engagement. Data were collected from 257 participants involved in construction projects, and the analysis was conducted using SPSS, AMOS, and the Hayes Process macro for mediation analysis. The results revealed that project governance is positively and significantly related to the project performance. While technological integration was found positive but insignificant with project performance, but their impacts was significantly enhanced when stakeholders engaged as mediator in the relationship. The mediation analysis underlined that stakeholder engagement is important for realizing governance structures and technological integration. Findings have important implications by expanding the understanding of the interplay between governance, technology, and human factors. This study offers theoretical and managerial implications.

Keywords: Project Governance · Technological Integration · Stakeholder Engagement · Project Performance · Agency Theory · Stakeholder Theory

1 Introduction

Megaprojects are large scale ventures that typically cost $1 billion or more, long term projects to build and develop, involving multiple public and private stakeholders, and resulting in to provide long term benefits for the people (Flyvbjerg 2021). Megaprojects are not the simplified version of small projects but they are completely different in all

E. Favari et al. (Eds.): MeRIT 2024, LNCE 600, pp. 55–71, 2025.
https://doi.org/10.1007/978-3-031-92797-3_5

the aspects, such as level of aspiration, lead times, stakeholder involvement, complexity and impact on the communities. Given that sustainable delivery is one of the most critical objectives in megaprojects. At each stage, the implementation of sustainable practices is essential for the success of these megaprojects (Guo and Zhang 2022). Project managers, policymakers, and stakeholders worldwide continue to prioritize megaproject performance. Megaprojects are frequently linked to cost overruns, schedule delays, and the failure to provide social and environmental benefits, despite their significance (Denicol and Davies 2022; Pitsis et al. 2018). These concerns have prompted inquiries regarding the governance mechanism and management practices of megaprojects, as well as the feasibility of such projects in achieving sustainability transition (Khalifeh et al. 2020). As environmental degradation and resource scarcity become more prevalent, the concept of sustainability has become more pertinent in the context of megaprojects (Ngampravatdee et al. 2023). The transition to sustainability refers to the transition from conventional, resource-intensive models of production and consumption to more sustainable, low-carbon alternatives (De Nito et al. 2024). Researchers are increasingly interested in the factors that influence the efficacy of megaprojects as their scope and complexity increase.

Project governance, as no one can deny its role in the success of projects, has emerged as a key to success, providing the frameworks and processes necessary to facilitate the decision makers, managing risks, and ensuring accountability (Müller et al. 2020). Effective governance mechanisms are essential for aligning the objectives of diverse stakeholders, mitigating conflicts, and ensuring that projects meet their intended goals. However, governance alone is not the single factor to guarantee the success. The rapid pace of technological advancement and the growing resilience on digital tools, and systems have also brought technological integration to the forefront of project management (Mijwil et al. 2023). Technological integration, defined as the seamless incorporation of technology into project processes, has the potential to enhance project performance by improving efficiency, reducing errors, and facilitating real-time decision making (Luo et al. 2023).

Similar to project governance, technological integration has emerged as a significant driver for the performance of megaprojects. The integration of digital tools, automation, and data-driven systems into project management processes is known as technological integration (Yang et al. 2021). These technologies facilitate real-time monitoring of project activities, predictive analytics for risk management, and improved communication among geographically dispersed teams. Megaprojects can enhance operational efficiency, reduce the probability of costly errors, and enhance decision-making by utilizing advanced technologies. Technologies such as Building Information Modeling (BIM), Internet of Things (IOT), and artificial intelligence (AI) have revolutionized how large projects can be controlled and managed (Geels et al. 2017). The successful integration of these technologies into megaproject workflows can deliver a substantial competitive edge, ensuring that project deadlines and performance indicators are achieved effectively. While both governance and technology are essential to the performance of megaprojects, the mechanisms by which these factors affect outcomes remain inadequately understood.

Stakeholders' involvement significantly contributes towards mediating the relationship of governance, technological integration and project performance during implementation of megaprojects. Stakeholder commitment is critical to addressing stakeholder needs and expectations. This means engagement of project managers, groups, and organizations that are stakeholders on the project (Adhi and Muslim 2023). Engaged stakeholders are more likely to provide valuable input, support project initiatives, and help resolve conflicts, thereby contributing to overall success of the project. According to Mashali and Eltantawy (2024), stakeholders engagement is particularly important in the context of megaprojects, where scale and complexity of operations often lead to a wide range of stakeholder interests, creating potential for misalignment. The complexity of megaprojects, which is a result of the involvement of multiple stakeholders, large budgets, extended timelines, and interdependent activities, is a significant challenge. The coordination challenges that frequently arise as a result of the projects' size make it challenging to guarantee that all components of the projects are in accordance with the objectives. Additionally, the potential for conflicts is exacerbated by the involvement of a diverse array of stakeholders, each with their own interests and priorities. To ensure the success of a project in such an environment, governance mechanisms that facilitate coordination and decision-making are essential. Simultaneously, it is imperative to implement technological instruments and implement systematic monitoring and communication to effectively manage complexities and guarantee the efficient execution of project activities.

Despite of the growing literature on project governance and technological integration, there remains a gap in understanding how these factors interact with stakeholder engagement to influence megaprojects performance.

While previous studies have explored the individual effects of project governance, technological integration, and stakeholder engagement on performance, little attention been paid to the mediating role of stakeholder engagement in these relationships. Iftikhar et al. (2021), looked into the crisis management, and studied about social, external and some internal technical issues in the megaprojects in Pakistan. But still the gap of how can we fill these crises in an effective way is missing. According to Bahadorestani et al. (2020) stakeholder engagement is not a static part of the project but rather a dynamic one that evolves throughout the project life cycle. According to Collinge (2020), effective engagement requires continuous communication, feedback loos, and the alignment of stakeholder interests with project goals. Especially in megaproject, where stakeholders interests are often diverse and conflicting, engagement becomes particularly challenging (Sankaran et al. 2020). This study provides an in-depth overview of the components that contribute to the success of megaprojects by illustrating the mediating function of stakeholder engagement. This framework is crucial for the attainment of sustainability because it guarantees that megaprojects are managed in a manner that prioritizes sustainable outcomes, employs technologies that facilitate efficiency and environmental responsibility, and actively engages stakeholders in every stage of the implementation process. The integrated approach enables us to more effectively achieve sustainability objectives, including the reduction of environmental impacts and the promotion of long-term social and economic benefits.

Especially, this research addresses three objectives:

1. To investigate the impact of project governance on project performance in megaprojects in Pakistan's construction industry.
2. To investigate how technology independently adds up in the project's performance.
3. To analyze the mediatory impact of stakeholder engagement in between project governance and technological integration relationship with the project performance (Fig. 1).

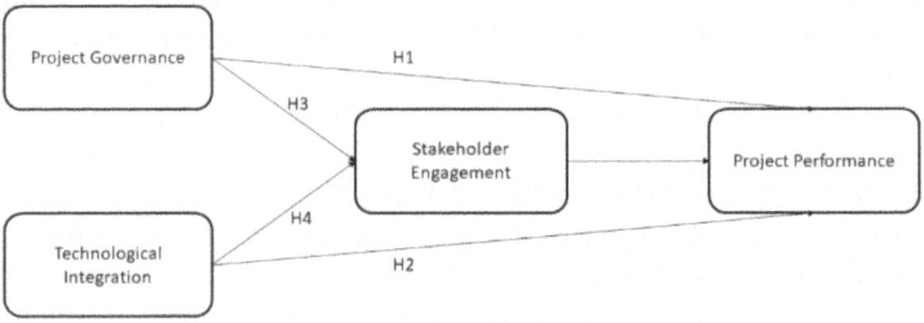

Fig. 1. Research Framework (Author's Made)

2 Theory and Hypotheses

Agency Theory and Stakeholder Theory offer a comprehensive framework for analyzing the dynamics of Project Governance, Technological Integration, and Stakeholder Engagement in megaprojects. Agency Theory Meckling and Jensen (1976), clarifies the principal-agent relationship, wherein the project owner (principal) confers decision-making authority to the project manager (agent), frequently resulting in knowledge asymmetry and divergent interests. This research argues that effective Project Governance is a means of addressing agency challenges by enhancing accountability, transparency, and alignment of incentives, thereby enhancing Project Performance. Stakeholder Theory Freeman and Phillips (2002), emphasizes the necessity of engaging all stakeholders in decision-making to harmonize varied interests with the project's goals. Within the framework of this study, Stakeholder Engagement serves as a mediator, facilitating the effective implementation of both governance and technical integration, hence augmenting total project success. Jointly, these concepts demonstrate how strategic alignment and collaboration among project stakeholders improve outcomes, particularly in the complex context of megaprojects.

2.1 Project Governance and Project Performance

Project governance and project performance has been extensively studied in the project management literature. Project governance involves the framework through which the decision making process, accountability, and control of all the measures are exercised within the project (Luo et al. 2024). In large-scale projects, the risk of misalignment, delays in decision-making, and cost overruns is heightened by the complexity and involvement of multiple stakeholders, making it imperative to ensure that projects meet their time, cost, and quality objectives (Dlamini and Cumberlege 2021).

Existing research emphasizes that project governance play a key role in risk management, decision making transparency, and stakeholder alignment, all of which are essential for achieving project success (Müller et al. 2017). However, previous studies highlight the importance of governance frameworks in improving project outcomes, the specific mechanism through which governance affects megaprojects performance remains underexplored. Given the unique challenges that megaprojects face, such as prolonged timelines, diverse stakeholders engagement, transparency in decision making, and following the standard regulation to meet objectives, the application of governance structures becomes even more important for success (Lovallo et al. 2023).

Hence, we posit that;

H1: Project governance has a positive and significant impact on Project performance.

2.2 Technological Integration and Project Performance

Technological integration has become the important factor in enhancing project performance, particularly in megaprojects. Technology and it's referring to the adoption of the seamless incorporation of the advanced tools, systems, and digital platforms. Such as Building Information Modeling (BIM), Internet of Things (IOT), and Artificial Intelligence (AI), into the project management processes to improve efficiency, collaboration, smooth flow of information, and decision making (Waqar et al. 2023). Celik et al. (2023), conducted research about these technologies, and he concludes that these technologies enable the real time-time monitoring, predictive analysis, and effective resource management, all of which contribute to the better performance of megaprojects.

Research has consistently demonstrated that technological integration can significantly improve project performance by enhancing communication, reducing errors, and enabling more informed decision-making (Leonard and Tyers 2023). However, while benefits of adopting technology in project management are well-documented, the specific mechanisms through which technological integration influences performance in megaprojects, those typically characterized by high complexity, long timelines, and the involvement of multiple stakeholders, remains unexplored. Previous studies often focused on small projects or single aspect of the technology, overlooking the comprehensive role that full technological integration plays improving the performance of megaprojects (Cantarelli and Genovese 2023). This suggests the gap in understanding how successful integration and utilization of technology, rather than more adoption, affects project outcomes, particularly in mega-construction projects, budget overruns, and coordination challenges are common (El-Sabek and McCabe 2018).

Furthermore, numerous projects are unable to fully maximize the potential of advanced technologies as a result of factors such as inadequate training for project teams, poor adoption, or a lack of stakeholder engagement (Sherman and Ford 2014). This implies a deficiency in comprehension regarding the impact of successful integration and utilization of technology on project outcomes, as opposed to mere adoption, particularly in megaprojects that frequently encounter coordination challenges, budget overruns, and delayed completion. (He et al. 2021). Therefore, the following hypothesis is proposed;

H2: Technological Integration has a positive and significant impact on Project performance.

2.3 Mediating Role Stakeholder Engagement

Stakeholder engagement refers to the process of involving individuals, groups, or organizations with a vested interest in the project throughput the project lifecycle. Effective engagement ensures that stakeholder needs, expectations, and concerns are adequately addressed, which can significantly influence project outcomes (Jakhar 2017). In the context of project management, the stakeholder engagement strategy can be referred as the salience level of stakeholders towards the projects. There are multiple strategies for engaging stakeholders, information sharing, involving them and maintaining gratification, adapting, collaborating, and proactive influencing (Dawkins 2014; Yasir and Farooq 2019). Megaprojects are often characterized by high level of complexity, uncertainty, and the involvement of diverse stakeholders with varying interests. This complexity necessitates not only formal governance structures and advanced technological tools but also the active participation and alignment of stakeholders to achieve the desired outcomes. We anticipate that the stakeholder engagement plays a role of mediator between project governance, technological integration and project performance. Mediation occurs exclusively when the mediator, a third variable, partially or entirely elucidates the relationship between the independent and dependent variables. The mediator provides a mechanism by which the independent variable influences the dependent variable (Baron and Kenny 1986). Research has shown that stakeholder engagement can act as a bridge that aligns the interest of all parties involved, leading to better project performance. According to Freeman's Stakeholder Theory (1984), the active involvement of stakeholders ensures that their needs and concerns are integrated into decision-making process., which ultimately leads to successful project outcomes.

Studies highlighted that stakeholder engagement serves as a mediator for fostering the collaboration and communication, which are essential for translating governance decisions and technological advancements into improved project outcomes (Bourne 2016). For example, when stakeholders are actively involved in governance processes, they are more likely to support and implement those decisions, thereby enhances the effectiveness of the governance framework in improving project success. Similarly, in the context of the technological context, engaged stakeholders are more likely to adopt and effectively use new technologies, thus amplifying the positive impact of these tools on project performance (Chen and Jin 2023; Ya and Rui 2006). Without this active engagement, project team members along with governance frameworks and technological innovations may remain underutilized or even resisted, leading to conflicts and inefficiencies and underperformance. Therefore, we posit that;

H3: Stakeholder engagement mediates the relationship between project governance and project performance.
H4: Stakeholder engagement mediates the relationship in between technological innovation and project performance.

3 Methodology

3.1 Research Design

The research follows a deductive approach based on the existing theories, including Agency theory and Stakeholder theory, to formulate the hypotheses that are then tested using empirical data. The study is cross sectional in nature. The data were collected from professionals involved in managing megaprojects in Pakistan. The quantitative approach is suitable for examining the strength and nature of the relationship between the key variables in the conceptual model and allow for robust statistical testing.

3.2 Sampling and Data Collection

For this study to examine the role of project governance and technological integration and megaproject performance with the mediation of stakeholder engagement, the data was gathered from Pakistan. Pakistan was chosen because the construction industry of Pakistan is in developing stage and it contributes of the 2.8% to the GDP of the total country. A purposive sampling technique was employed to ensure the respondents to have significant experience with megaprojects and can provide the relevant insight into study objective. A sample size of 257 respondents was chosen to achieve the statistical significance of our data. To test our hypothesis we used SPSS 23, and AMOS 21 for model fit analysis.

3.3 Measurement of Variables

Project Governance (PG) is our first independent variable. This variable is measured using items to access the presence of governance structure, decision-making framework, risk management, and accountability mechanism. Items were adapted from Müller and Turner (2007). A total of 4-items were chosen, and rated on 5 items Likert scale ranging from "Strongly Agree to Strongly Disagree". Sample items include "The project governance structure ensures compliance with relevant standards and regulations". ($\alpha = 0.703$, AVE = 0.71, CR = 0.83).

Technological Integration (TI) is our second independent variable. This research is contingent to access how level of adoption and effective use of technologies such as BIM, IOT, and AI within the project management processes adds to the project. Items were adapted from Reichstein et al. (2005). 4 Items were chosen, and rated on 5 items Likert scale ranging from "Strongly Agree to Strongly Disagree". Sample items include "The project utilizes advanced technologies to enhance performance". ($\alpha = 0.781$, AVE = 0.78, CR = 0.89).

Stakeholder Engagement (SE) is our mediating variable. The study items helped us to evaluate how well stakeholders are involved in project decision-making, collaboration, and conflict resolution. Items are adapted from Yang et al. (2009). 4 items were chosen, and rated on 5 items Likert scale ranging from "Strongly Agree to Strongly Disagree". Sample items include "Stakeholders are actively involved in the decision-making process". ($\alpha = 0.828$, AVE = 0.76, CR = 0.84).

Project Performance (PP) is our dependent variable. The project performance dimensions helped us to measure the time, cost, quality, and stakeholder's satisfaction. Items were adapted from Khalifeh et al. (2020). 5 items were chosen and rated on 5 items Likert scale ranging from "Strongly Agree to Strongly Disagree". Sample items include "The quality of megaproject achieves the expected standard". ($\alpha = 0.764$, AVE = 0.75, CR = 0.80).

4 Results

Initially, we conducted a series of confirmatory factor analyses (CFA) to validate the goodness-of-fit of the measurement model utilizing AMOS version 21. All CFAs' results are displayed in Table 1. The goodness of fit for all CFAs validates the discriminant validity of the specified variables. We employed Cronbach's alpha (α) values to assess the internal consistency of the specified constructs. The findings in Table 2 indicate that all α values exceeded the 0.75 criterion, hence confirming the internal consistency of the constructs (Nunnally 1978). We evaluated the standardized factor loadings, composite reliability (CR), and average variance extracted (AVE) to determine the convergent validity of the instrument.

Table 1. Confirmatory Factor Analysis (CFA)

Model(s)	Chi 2	Df	CFI	GFI	TLI	RMR	RMSEA
2 factor (PG + PP)	55.590	26	.950	.956	.930	.030	.067
2 factor (TI + PP)	51.502	26	.963	.957	.949	.027	.062
3 factor (PG + PP+ SE)	113.034	56	0654	.938	.937	.030	.063
3 factor (TI + PP+ SE)	124.924	55	.952	.934	.932	.029	.070
4 factor (PG + TI+ SE+ PP)	241.788	107	.927	.900	.907	.029	.070

Table 2. Descriptive, Correlation and Discriminant Validity Results

Construct	Mean	SD	1	2	3	4
1. Project governance	4.17	0.49	**(0.971)**			
2. Technological integration	4.23	0.55	0.624**	**(0.920)**		
3. Stakeholder engagement	4.25	0.58	0.558**	0.690**	**(0.893)**	
4. Project performance	3.89	0.56	0.346**	0.346**	0.392**	**(0.877)**

Note(s): N = 257. *$p < 0.05$, **$p < 0.01$

Finally, we assessed the discriminant validity of the constructs by confirming that the square root of the AVE exceeded the correlations of the specified variables with all other variables (Fornell and Larcker 1981). The values in Table 4 indicate that the conditions for discriminant validity have been met, as the square root of the AVE exceeds the correlations of each respective construct.

4.1 Testing of Hypotheses

Linear Regression Results

Model 2 results in Table 3, indicate a positive and significant relationship between Project Governance (β = 0.174, p = 0.040) and Project Performance. This finding corroborates the initial hypothesis (H1) and indicates that enhanced governance systems, such as explicit decision-making frameworks, accountability, and risk management, improve project outcomes. The positive coefficient signifies those enhancements in project governance correlate with improved project performance, underscoring the essential function of governance frameworks in navigating the complexities of megaprojects. The association between project governance and project performance is statistically significant at a level below 0.05, affirming its relevance in large-scale initiatives.

Table 3. Linear Regression Results

Variables	Model 1 Coeff.	Std. Error	Prob	Model 2 Coeff.	Std. Error	Prob
Control variables						
Project complexity	−.137	.118	0.011	−0.091	0.050	0.066
Main effects						
Project governance				0.174	0.084	0.040
Technological INTG.				0.76	0.84	0.387
St. engagement				0.232	0.077	0.003
R^2	0.025			0.190		
Adj. R^2	0.021			0.177		
F-statistics	6.540**			14.788***		

In Model 2, Table 3, the correlation between Technological Integration and Project Performance is favorable however statistically insignificant (β = 0.76, p = 0.387). This outcome indicates that although the incorporation of modern technology into project management procedures may positively impact performance, the effect is insufficiently robust to be deemed statistically significant in this model. The non-significant p-value suggests that technological integration alone may be inadequate for enhancing project performance, potentially necessitating additional elements such as effective stakeholder involvement or complementing governance frameworks to fully achieve its advantages. Consequently, the second hypothesis (H2) is not supported by this research.

Regression Analysis for Mediation

Mediation analysis in Table 4, indicate that Stakeholder Engagement serves as a mediator between Project Governance and Project Performance. The total effect of Project Governance (PG) on Project Performance (PP) is statistically significant (Effect = 0.3595, SE = 0.0662, t = 5.8850, $p < 0.001$). This suggests that governance has a substantial positive influence on performance when no mediator is taken into account.

Table 4. Mediation Analysis (PG – SE – PP) Process Model 4

Total Effects					95% BCa CI	
	Effect	**SE**	**T**	**P**	**LB**	**UB**
PG → PP	.3595	.0662	5.8850	.0000	.2592	.5199
Direct effect					**95% BCa CI**	
	Coefficient	**SE**	**t-value**	**P**	**LB**	**UB**
PG → PP	.2083	.0773	2.6958	.0075	.0561	.3604
Indirect effects			**95% BCa CI**			
	Effect	**BootSE**	**BootLLCI**	**BootULCI**		
PG → SE → PP	.1813	.0561	.0782	.3015		

Despite the inclusion of Stakeholder Engagement (SE) as a mediator, the direct impact of Project Governance on Project Performance remains substantial (Coefficient = 0.2083, SE = 0.0773, t = 2.6958, p = 0.0075). This suggests that project governance has a direct positive impact on project performance, although the effect magnitude is reduced when the mediator is taken into account. As the confidence interval does not include zero, the indirect effect through Stakeholder Engagement is also significant (Effect = 0.1813, BootSE = 0.0561, 95% CI [0.0782, 0.3015]). This suggests that Stakeholder Engagement partially mediates the relationship between project governance and project performance.

Therefore, the mediation analysis confirms that Stakeholder Engagement significantly enhances the impact of Project Governance on Project Performance, thereby supporting the hypothesis 3, that stakeholder involvement confirms the positive influence of project governance on project outcomes.

The results of the mediation analysis in Table 5 indicate that Stakeholder Engagement mediates the relationship between Technological Integration (TI) and Project Performance (PP). Technological integration has a substantial positive impact on project performance when the mediator is excluded (Effect = 0.3537, SE = 0.0601, t = 5.8897, $p < 0.001$).

Nevertheless, the direct impact of Technological Integration on Project Performance is rendered insignificant when Stakeholder Engagement is incorporated as a mediator (Coefficient = 0.1479, SE = 0.0810, t = 1.8261, p = 0.0690). The direct effect's confidence interval (CI) is zero (−0.0016 to 0.3075), suggesting that the direct relationship between project performance and technological integration is not statistically significant. The confidence interval does not include zero, which is why the indirect effect through Stakeholder Engagement is significant (Effect = 0.2058, BootSE = 0.0734, 95% CI [0.0797, 0.3710]). This suggests that the relationship between Technological Integration and Project Performance is wholly mediated by Stakeholder Engagement.

Consequently, the mediation analysis confirms that Stakeholder Engagement is a critical factor in the transformation of the positive impact of Technological Integration into

Table 5. Mediation Analysis (TI – SE – PP) Process Model 4

Total Effects					95% BCa CI	
	Effect	**SE**	**T**	**P**	**LB**	**UB**
TI → PP	.3537	.0601	5.8897	0.0000	.2355	.4720
Direct effect					95% BCa CI	
	Coefficient	**SE**	**t-value**	**P**	**LB**	**UB**
TI → PP	.1479	.0810	1.8261	.0690	−.0016	.3075
Indirect effects			95% BCa CI			
	Effect	**BootSE**	**BootLLCI**	**BootULCI**		
TI → SE → PP	.2058	.0734	.0797	.3710		

improved Project Performance, indicating that the success of technological integration is contingent upon the effectiveness of stakeholder engagement. Hence our hypothesis is supported.

5 Discussion of Results

First hypothesis asserts that Project Governance has a beneficial impact on Project Performance. In megaprojects, where complexity, scope, and multiple stakeholders can result in substantial risks, governance is a critical determinant of project success (Müller 2019). Effective governance creates the necessary structures for decision-making, accountability, and risk management to align project objectives with stakeholder expectations (Too and Weaver 2014). Project governance serves as a framework to ensure compliance, coordination, and accountability in megaprojects, where mismanagement and misalignment of objectives may lead to delays, cost overruns, and failures (Flyvbjerg 2021). Governance techniques, including clearly delineated roles and duties, transparent decision-making processes, and efficient risk management, proved essential for achieving successful project outcomes. Participants who evaluated these governance elements favorably were more inclined to indicate that their initiatives adhered to the established scope, budget, and quality criteria. Empirical research indicates that projects that are well-governed exhibit superior performance outcomes in terms of meeting their schedule, cost, and quality objectives (Denicol et al. 2020). Thus, this hypothesis is supported by the existing governance literature, which highlights its importance in improving project performance through the promotion of accountability, the enhancement of communication, and the reduction of risks in extensive project settings.

Hypothesis 2, posits that Technological Integration positively influences Project Performance. In the context of digital transformation, the incorporation of advanced technologies like Building Information Modeling (BIM), Internet of Things (IoT), and Artificial Intelligence (AI) into project management is increasingly regarded as essential for improving efficiency, coordination, and decision-making (Guo and Zhang 2022). The utilization of these technologies in megaprojects enhances operational efficiency, facilitates real-time progress monitoring, and optimizes resource allocation, hence substantially improving project performance (Geels et al. 2023; Sirisomboonsuk et al. 2018; Terenzi et al. 2024). The utilization of advanced technologies and their capacity for seamless integration into project workflows underscores the necessity of sufficient training and support. The negligible direct influence of technological integration on performance indicates that, although technology can improve project results, its effectiveness primarily relies on the proficiency of the project team in adopting and utilizing it. Technologies lacking adequate integration or training support may not achieve their intended advantages. Nevertheless, the literature emphasizes that mere technology integration is inadequate; its efficacy predominantly relies on the extent to which the technologies are embraced and employed inside the project management framework (Ghosh et al. 2017). Recent research indicates that projects employing digital tools demonstrate significant enhancements in outcomes, especially regarding cost and time efficiency (Li et al. 2019). This hypothesis is based on the premise that effective technology integration enhances efficiency and overall performance in complicated projects.

Hypothesis 3, stated that Stakeholder Engagement mediates the relationship between Project Governance and Project Performance. According to Eskerod and Larsen (2018), efficacy of governance structures, which establish formal frameworks for accountability and decision-making, is frequently dependent on the degree of stakeholder engagement. Stakeholder engagement is the active participation of important stakeholders in project decision-making, thereby ensuring alignment between their interests and the project's objectives. In megaprojects, when various stakeholders with differing objectives are engaged, participation is essential for conflict resolution, improving collaboration, and assuring the effective implementation of governance systems (Luyet et al. 2012). The sample results indicates that projects with greater stakeholder participation are more likely to succeed, highlighting the importance of active involvement, effective communication, and dispute resolution. Stakeholders engaged in decision-making and content with their participation are more likely to endorse the project's objectives, hence improving overall performance. Empirical research indicates that active stakeholder engagement enhances adherence to governance systems, resulting in increased project performance (Mok et al. 2015). This hypothesis underscores the essential function of stakeholder engagement in amplifying the beneficial impacts of governance on project results, especially in extensive, intricate projects where stakeholder alignment is crucial for success.

Hypothesis 4, Although the direct relationship of technological integration with project performance was not fully significant. But when the stakeholder engagement mediates in between them, it is accepted. This intricates the role of social and collaborative processes in leveraging the technological advancements mutually. It enables the projects with all the particular tools and systems those necessary for maintaining the efficient data sharing, real time monitoring, and prediction analytics (Rao et al. 2019). However, the full benefits of these technologies can be multiplied with the active role of stakeholders in the project life cycle.

According to Shah and Guild (2022), who's research looks critically to the stakeholders engagement and further they declared that involvement of stakeholders in the early stages of the projects can be more beneficial as vice versa. For instance, the implementation of IOT or digital twin technologies can improve the communication, fostering trust, and facilitating shared decision making before going into any misunderstandings. This entrusts, the projects outcomes reducing unnecessary delays, mitigating conflicts, and aligning project objectives with overall stakeholders' interests into the whole project life cycle. Study findings were found consistent with the earlier literature as, Nguyen and Mohamed (2021), concluding that stakeholder management significantly enhances the project outcomes by aligning the diverse interests. Zhang et al. (2023), concluded that during BIM implementation role of stakeholder was found positively towards the overall project performance.

5.1 Theoretical Implications

This research enhances the current understanding of project management by synthesizing Agency Theory and Stakeholder Theory to make clear the complex dynamics of Project Governance, Technological Integration, and Stakeholder Engagement in influencing Project Performance in mega-construction projects. The results indicate that stakeholder engagement mediates the relationship among governance, technology, and project outcomes, elucidating the necessity for alignment between governance frameworks and technological instruments with stakeholder interests to attain effective project results. This study enhances current ideas by highlighting the essential importance of active stakeholder engagement in megaprojects, when governance and technology alone are inadequate for achieving performance. The study emphasizes the importance of integrating sustainability goals into governance and stakeholder frameworks, hence enhancing the ongoing discussion on sustainability in project management. This study enhances the theoretical comprehension of how collaboration and communication among project participants influence project success, particularly in complex and large-scale projects, by including stakeholder involvement as a mediating variable.

5.2 Managerial Implications

The conclusions of this research are particularly pertinent to the construction sector of Pakistan. The construction sector in Pakistan is marked by the implementation of significant infrastructure projects, typically financed by both local and international investors. The China-Pakistan Economic Corridor (CPEC) programs encompass a multifaceted network of partners, comprising governmental bodies, commercial contractors, international investors, and local people. The construction industry in Pakistan has historically encountered obstacles stemming from ineffective governance, inadequate stakeholder participation, and technology deficiencies, resulting in delays, cost overruns, and quality concerns. Considering the magnitude and intricacy of such megaprojects, the study's focus on stakeholder participation and efficient governance is particularly relevant for enhancing project outcomes in Pakistan.

Project managers in Pakistan can derive numerous useful insights from this study from a managerial standpoint. In large-scale construction projects, it is essential for managers to build strong governance frameworks that clarify responsibilities, decision-making, and risk management, while also engaging stakeholders actively. The parties are regularly engaged in construction projects, it is essential to consult stakeholders from governmental bodies to local communities to harmonize interests, mitigate tensions, and maintain project momentum. Secondly, the technological integration within Pakistan's building sector frequently falls short of worldwide benchmarks. The study emphasizes that the implementation of modern technologies, such as Building Information Modeling (BIM) or automated project management tools, can improve project performance, contingent upon the proper training and engagement of stakeholders in the process. Managers must consequently invest in capacity building to enable all stakeholders to properly utilize new technologies, resulting in enhanced efficiency, budget compliance, and sustainability in project outputs. By implementing these managerial approaches, the Pakistani construction sector may surmount prevalent problems and improve the overall success of its megaprojects.

5.3 Conclusion and Limitation of This Research

This study offers an in-depth examination of the roles of Project Governance, Technological Integration, and Stakeholder Engagement in enhancing Project Performance in large-scale building endeavors, especially in the realm of megaprojects. The study presents a refined comprehension of the efficacy of governance frameworks and sophisticated technical solutions when facilitated by proactive and involved stakeholders. The results highlight that simply implementing governance frameworks or using advanced technologies is inadequate for ensuring project success. The active involvement, communication, and collaboration of stakeholders greatly enhance the beneficial effects of these aspects on project results. This study enhances the project management literature by affirming that Stakeholder Engagement is a fundamental element that mediates and improves the efficacy of governance and technology in realizing sustainable and high-performing megaprojects. This research highlights the significance of comprehensive project management strategies that incorporate structural, technological, and human elements to attain optimal project outcomes.

This research provides useful insights; nonetheless, its applicability in the Pakistan's construction industry has specific constraints. Construction businesses, while experiencing tremendous growth, faces numerous systemic obstacles, such as political instability, variable laws, and inconsistent enforcement of standards. These circumstances may hinder the establishment of effective governance frameworks and restrict the comprehensive incorporation of modern technology. Moreover, stakeholder participation in Pakistan frequently encounters cultural and logistical obstacles, resulting in communication deficiencies among project managers, contractors, local populations, and international investors. This may impede the efficacy of stakeholder participation in decision-making and conflict resolution. Moreover, although the research highlights the beneficial effects of technology integration, the availability and accessibility of these technologies in Pakistan may be hindered by infrastructural deficiencies, skill shortages, and financial limitations. Consequently, while the findings are relevant, their effective application in the Pakistani construction sector necessitates surmounting specific local problems that this research does not comprehensively address. It is advisable to do additional research focused on the socio-political and economic conditions of Pakistan to formulate more context-specific methods for enhancing megaproject success.

References

1. Adhi AB, Muslim F (2023) Development of stakeholder engagement strategies to improve sustainable construction implementation based on lean construction principles in Indonesia. Sustain For 15(7):6053
2. Bahadorestani A, Naderpajouh N, Sadiq R (2020) Planning for sustainable stakeholder engagement based on the assessment of conflicting interests in projects. J Clean Prod 242:118402
3. Baron RM, Kenny DA (1986) The moderator–mediator variable distinction in social psychological research: conceptual, strategic, and statistical considerations. J Pers Soc Psychol 51(6):1173–1182
4. Bourne L (2016) Stakeholder relationship management: a maturity model for organisational implementation. Routledge
5. Cantarelli CC, Genovese A (2023) Innovation potential of megaprojects: a systematic literature review. Prod Plan Control 34(14):1350–1370
6. Celik Y, Petri I, Barati M (2023) Blockchain supported BIM data provenance for construction projects. Comput Ind 144:103768
7. Chen Y, Jin S (2023) Corporate social responsibility and green technology innovation: the moderating role of stakeholders. Sustain For 15(10):8164
8. Collinge W (2020) Stakeholder engagement in construction: exploring corporate social responsibility, ethical behaviors, and practices. J Constr Eng Manag 146(3):04020003
9. Dawkins CE (2014) The principle of good faith: toward substantive stakeholder engagement. J Bus Ethics 121:283–295
10. De Nito E, Caccialanza A, Canonico P, Favari E (2024) Analyzing the role of social value in megaprojects: toward a new performance framework. Meas Bus Excell 28(2):243–254
11. Denicol J, Davies A (2022) The megaproject-based firm: building programme management capability to deliver megaprojects. Int J Proj Manag 40(5):505–516
12. Denicol J, Davies A, Krystallis I (2020) What are the causes and cures of poor megaproject performance? A systematic literature review and research agenda. Proj Manag J 51(3):328–345
13. Dlamini M, Cumberlege R (2021) The impact of cost overruns and delays in the construction business. IOP Conf Series Earth Environ Sci 654:012029
14. El-Sabek LM, McCabe BY (2018) Coordination challenges of production planning in the construction of international mega-projects in the Middle East. Int J Constr Educ Res 14(2):118–140
15. Eskerod P, Larsen T (2018) Advancing project stakeholder analysis by the concept 'shadows of the context'. Int J Proj Manag 36(1):161–169
16. Flyvbjerg B (2021) Top ten behavioral biases in project management: An overview. Proj Manag J 52(6):531–546
17. Fornell C, Larcker DF (1981) Evaluating structural equation models with unobservable variables and measurement error. J Mark Res 18(1):39–50
18. Freeman RE, Phillips RA (2002) Stakeholder theory: a libertarian defense. Bus Ethics Q 12(3):331–349
19. Geels FW, Sovacool BK, Iskandarova M (2023) The socio-technical dynamics of net-zero industrial megaprojects: outside-in and inside-out analyses of the Humber industrial cluster. Energy Res Soc Sci 98:103003
20. Geels FW, Sovacool BK, Schwanen T, Sorrell S (2017) Sociotechnical transitions for deep decarbonization. Science 357(6357):1242–1244
21. Ghosh M, Kabir G, Hasin MAA (2017) Project time–cost trade-off: a Bayesian approach to update project time and cost estimates. Int J Manag Sci Eng Manag 12(3):206–215

22. Guo K, Zhang L (2022) Multi-objective optimization for improved project management: current status and future directions. Autom Constr 139:104256
23. He Q, Xu J, Wang T, Chan AP (2021) Identifying the driving factors of successful megaproject construction management: findings from three Chinese cases. Front Eng Manag 8(1):5–16
24. Iftikhar R, Müller R, Ahola T (2021) Crises and coping strategies in megaprojects: the case of the Islamabad–Rawalpindi Metro Bus Project in Pakistan. Proj Manag J 52(4):394–409
25. Jakhar SK (2017) Stakeholder engagement and environmental practice adoption: the mediating role of process management practices. Sustain Dev 25(1):92–110
26. Khalifeh A, Farrell P, Al-edenat M (2020) The impact of project sustainability management (PSM) on project success: a systematic literature review. J Manag Dev 39(4):453–474
27. Leonard P, Tyers R (2023) Engineering the revolution? Imagining the role of new digital technologies in infrastructure work futures. N Technol Work Employ 38(2):291–310
28. Li Y, Han Y, Luo M, Zhang Y (2019) Impact of megaproject governance on project performance: dynamic governance of the Nanning transportation hub in China. J Manag Eng 35(3):05019002
29. Lovallo D, Cristofaro M, Flyvbjerg B (2023) Governing large projects: a three-stage process to get it right. Acad Manag Perspect 37(2):138–156
30. Luo L, Liu Y, Yang Y, Xie J, Wu G (2024) Interactive roles of megaproject governance mechanisms and their effects on governance performance. In: Engineering, construction and architectural management
31. Luo Y, Lu Z, Wu C (2023) Can internet development accelerate the green innovation efficiency convergence: evidence from China. Technol Forecast Soc Chang 189:122352
32. Luyet V, Schlaepfer R, Parlange MB, Buttler A (2012) A framework to implement stakeholder participation in environmental projects. J Environ Manag 111:213–219
33. Mashali A, Eltantawy A (2024) Practical approach for analysing and engaging stakeholders in construction megaprojects. Build Eng 2(1):509–509
34. Meckling WH, Jensen MC (1976) Theory of the firm. In: Managerial behavior, agency costs and ownership structure
35. Mijwil MM, Hiran KK, Doshi R, Unogwu OJ (2023) Advancing construction with IoT and RFID technology in civil engineering: a technology review. Al-Salam J Eng Technol 2(02):54–62
36. Mok KY, Shen GQ, Yang J (2015) Stakeholder management studies in mega construction projects: a review and future directions. Int J Proj Manag 33(2):446–457
37. Müller R (2019) The relationship between project governance and project success. In: Project management methodologies, governance and success. Auerbach Publications, pp 141–170
38. Müller R, Drouin N, Sankaran S (2020) Governance of organizational project management and megaprojects using the viable project governance model. In: Handbook of systems sciences. Springer, Singapore, pp 1–27
39. Müller R, Turner R (2007) The influence of project managers on project success criteria and project success by type of project. Eur Manag J 25(4):298–309
40. Müller R, Zhai L, Wang A (2017) Governance and governmentality in projects: profiles and relationships with success. Int J Proj Manag 35(3):378–392
41. Ngampravatdee C, Gharehbaghi K, Hosseinian-Far A, Tee KF, McManus K (2023) Strategic initiatives for large transport infrastructure planning: reinforcing sustainability in urban transportation through better stakeholder engagement. Sustain For 15(18):13912
42. Nguyen TS, Mohamed S (2021) Mediation effect of stakeholder management between stakeholder characteristics and project performance. J Eng Proj Prod Manag 11(2)
43. Nunnally JC (1978) An overview of psychological measurement. In: Clinical diagnosis of mental disorders: A handbook, pp 97–146
44. Pitsis A, Clegg S, Freeder D, Sankaran S, Burdon S (2018) Megaprojects redefined–complexity vs cost and social imperatives. Int J Manag Proj Bus 11(1):7–34

45. Rao TR, Mitra P, Bhatt R, Goswami A (2019) The big data system, components, tools, and technologies: a survey. Knowl Inf Syst 60:1165–1245
46. Reichstein T, Salter AJ, Gann DM (2005) Last among equals: a comparison of innovation in construction, services and manufacturing in the UK. Constr Manag Econ 23(6):631–644
47. Sankaran S, Müller R, Drouin N (2020) Creating a 'sustainability sublime' to enable megaprojects to meet the United Nations sustainable development goals. Syst Res Behav Sci 37(5):813–826
48. Shah MU, Guild PD (2022) Stakeholder engagement strategy of technology firms: a review and applied view of stakeholder theory. Technovation 114:102460
49. Sherman MH, Ford J (2014) Stakeholder engagement in adaptation interventions: an evaluation of projects in developing nations. Clim Pol 14(3):417–441
50. Sirisomboonsuk P, Gu VC, Cao RQ, Burns JR (2018) Relationships between project governance and information technology governance and their impact on project performance. Int J Proj Manag 36(2):287–300
51. Terenzi M, Locatelli G, Winch GM (2024) Projects as vectors of change: a transition toward net-zero sociotechnical systems. Proj Manag J:87569728241270578
52. Too EG, Weaver P (2014) The management of project management: a conceptual framework for project governance. Int J Proj Manag 32(8):1382–1394
53. Waqar A, Khan MB, Shafiq N, Skrzypkowski K, Zagórski K, Zagórska A (2023) Assessment of challenges to the adoption of IOT for the safety management of small construction projects in Malaysia: structural equation modeling approach. Appl Sci 13(5):3340
54. Ya S, Rui T (2006) The influence of stakeholders on technology innovation: a case study from China. In: 2006 IEEE international conference on management of innovation and technology
55. Yang J, Shen GQ, Ho M, Drew DS, Chan AP (2009) Exploring critical success factors for stakeholder management in construction projects. J Civ Eng Manag 15(4):337–348
56. Yang L, Huo B, Tian M, Han Z (2021) The impact of digitalization and inter-organizational technological activities on supplier opportunism: the moderating role of relational ties. Int J Oper Prod Manag 41(7):1085–1118
57. Yasir M, Farooq S (2019) Project success, internal stakeholder engagement and satisfaction: a moderated mediation analysis. J Human Soc Sci 27(2):29–56
58. Zhang HM, Chong H-Y, Zeng Y, Zhang W (2023) The effective mediating role of stakeholder management in the relationship between BIM implementation and project performance. Eng Constr Archit Manag 30(6):2503–2522

Legal and Regulatory Foundations for Megaprojects Execution

The Reform of the Italian Tax System and Its Effect on the Implementation of Megaprojects

Marco Allena(✉)

Università Cattolica del Sacro Cuore, Milan, Italy
marco.allena@unicatt.it

Abstract. The paper identifies links between the recent reform of the Italian taxation system, which began with the tax reform enabling law no. 111 of 9 August 2023, and megaprojects. Particular attention is paid to the effects that an initial evaluation suggests have been triggered for entities intending to commit to megaprojects following reform of the tax residence of legal entities, changes to the rules governing the controlled foreign company (CFC) tax regime, introduction of the global minimum tax into the Italian legal system, the anticipated application of measures to incentivize reshoring and, finally, changes to the cooperative compliance regime.

1 Introduction

The international literature defines megaprojects as complex, large-scale investments in a variety of public and private sectors, with an investment value in excess of USD one million or EUR one million. Such projects have a considerable level of associated risk, are very difficult to manage, and the final result often differs greatly from initial expectations.

There are a number of variables that can affect the execution of megaprojects. It is not unusual for them to be completed in a manner and for costs that are significantly different from those originally predicted.

Furthermore, given the complexity and level of innovation in megaprojects, it is not easy to plan them with absolute certainty. They are unique, and different from any other project. To put it another way, when a project is placed within a defined context it is relatively easy to study the reaction of the surrounding environment using the predictive models available, whereas the same cannot be said of a work or complex that significantly alters the eco-system.

Obviously, the risk profiles that need to be taken into account by those involved in megaprojects include the taxation element, i.e. the tax laws affecting anyone deciding to invest in and launch a megaproject in a given country.

Therefore, knowledge of tax regulations, tax certainty and a preventive dialogue with the tax authorities of the country in question in order to find out about the applicable tax regimes, where possible, are an absolute pre-requisite for those operating in the field of megaprojects.

E. Favari et al. (Eds.): MeRIT 2024, LNCE 600, pp. 75–86, 2025.
https://doi.org/10.1007/978-3-031-92797-3_6

With this in mind, below we identify links between megaprojects and the reform of the Italian taxation system that began with the tax reform enabling law no. 111 of 9 August 2023 and has been implemented by subsequent enabling legislative decrees.

More specifically, we highlight the effects that the ongoing reform of the Italian taxation system could have on megaprojects currently under way and, in particular, those that could be developed in the immediate future.

We focus our attention on some areas of the reform which seem to have particular relevance for megaprojects.

These are:

- Reform of the tax residence of legal persons
- Introduction of the global minimum tax (GMT)
- The new controlled foreign company (CFC) rules
- New reshoring measures
- The new cooperative compliance regime.

2 Reform of the Tax Residence of Legal Persons. What Are the Effects on Megaprojects?

The scope of the tax residency of legal persons has been profoundly impacted by Legislative Decree no. 209 of 27 December 2023, which enacted enabling law no. 111 of 9 August 2023.

The reform is not insignificant and, as shown below, has clear and important repercussions for those involved in megaprojects. This is because the choice of a given tax jurisdiction based on a company's tax residence implies the application of specific tax regulations, adoption of accounting procedures mandatory in a given tax system and compliance with the terms and conditions the latter applies to the filing of tax returns.

It is evident that reforming the criteria for determining the residence of legal persons could have an impact on entities currently involved or planning to be involved in megaprojects. Not least because such criteria affect the taxation, costs, incentives and tax planning of entities operating in megaprojects.

In other words, companies involved in the implementation of major works must consider carefully where to have their tax residence. The decision has effectively become part of their investment and management strategy, the goals of which are to maximize benefits and minimize risks.

On this point, it is important to note that, with effect from 1 January 2024, article 2 of Legislative Decree no. 209 of 2023 amended article 73 of the TUIR (consolidated income tax law), which affects those subject to income tax. The decree amended the third paragraph of article 73 on the tax residence of corporate income tax (IRES) payers.

According to the previous formula, companies and entities were tax resident in Italy if their registered office, corporate headquarters or main corporate purpose was located in the country for the majority of the financial year (equal to or more than 183 days or 184 days in a leap year).

Article 2 of the decree has not touched the linking criterion relating to the presence of the company or entity's registered office in the country.

However, it has removed the main purpose criterion and reformulated the corporate headquarters criterion by identifying the following substantive linking criteria: effective direction and principal day-to-day management.

Given the new formulation of article 73 of the TUIR, the place of effective direction means the continuous and coordinated taking of strategic decisions concerning the company or entity as a whole. Similarly, day-to-day management means the continuous and coordinated management of the company or entity as a whole.

Effective direction does not include decisions other than those with management content taken by shareholders, nor does it include the supervision or monitoring of management by the same. Conversely, management activities by the shareholder that interfere with the typical functions of the management body of the controlled company continue to be important.

An initial evaluation would suggest that adoption of these criteria would not only respond to the need for greater consistency with international law, but also guarantee greater legal certainty to taxpayers in defining the parameters for recognising tax residence.

Therefore, as a result of the reform, the new tax residence according to article 73 of the TUIR is now determined on the basis of three alternative criteria, of which one – the registered office criterion – of a formal nature, and the other two – the criteria of the place of effective direction and principal day-to-day management – of a substantive nature.

As the literature shows, the fact that the lawmakers have introduced the criteria of the place of effective direction and day-to-day management to replace corporate headquarters, at the same time eliminating the criterion of main purpose, has a two-fold objective:

- To reinforce legal certainty and overcome the difficulties of interpretation and application that had progressively weakened the indicative purpose of the corporate headquarters criterion and made understanding the main purpose requisite less straightforward, leading to a significant number of disputes;
- To eliminate the root causes of improper expansion to include additional criteria of a substantive nature that could progressively distort the content and function of the concept of residence.

From this viewpoint, reform of the residence criteria for legal persons must be welcomed also by those operating in the area of megaprojects. Not least because it reduces the potential for disputes with the tax authorities. The lawmakers have specifically aimed to add content to the criterion of the place of direction precisely in order to avoid arbitrary interpretations.

In fact, where the rule now specifies that the place of direction must be understood to mean "*the continuous and coordinated taking of strategic decisions concerning the company or entity as a whole*", it clearly delimits the scope of the criterion, clearing the field of possible extensions of interpretation which, when the previous version of article 73 was in force, had weakened the boundaries of the notion of corporate headquarters.

Similarly, the introduction of the criterion of principal day-to-day management should also be welcomed, especially with respect to the criterion of "main purpose" which has been eliminated, as its application had engendered several issues at an operational level.

As the literature shows, as for the "place of effective direction" criterion, the lawmakers chose a solution that would define the specific content of the new criterion, in order to guide and simplify its interpretation.

An initial evaluation would suggest that these legislative amendments are beneficial for the world of business and therefore also for companies involves in megaprojects.

Given the legislative framework, those planning to implement megaprojects in Italy, or involved in their implementation, could benefit from more certain rules concerning the tax residence of their company.

The rules appear to produce at least two effects.

The first is to move away from the uncertainty surrounding application of the previous main purpose criterion, especially as regards identifying the residence of companies – and therefore, more specifically, of those operating in megaprojects – as being the country in which the company's assets are principally located (rather than in the place where the company's business and negotiations are carried out in order to achieve the company purpose).

The second is to abandon objections essentially based on equating the company's control structure with that of the top management.

In fact, following the reform, shareholders' actions could become important for the purpose of confirming the place of effective direction of a company working in megaprojects only in the event that such actions involve taking decisions of a managerial nature and therefore when shareholders are also directors. In other words, where they actually have the functions and powers of directors.

3 Application of the Global Minimum Tax (GMT) to Megaprojects

As mentioned above, the implementation of a megaproject often involves national and multinational corporate groups that also operate outside of strictly domestic markets. Clearly megaproject planning has to take the rules governing the taxation of the income generated by large corporate groups into due consideration. This is particularly in light of the recent international, European and finally national introduction of a global minimum tax (GMT), the main purpose of which is to reduce erosion of the taxable base and the shifting of profits by multinationals to more favourable tax jurisdictions.

Application of the tax comes at the end of a long and complex process launched with the OECD's BEPS (Base Erosion and Profit Shifting) project, culminating in Europe with Council Directive (EU) 2022/2523 of 14 December 2022, which aimed to ensure a global minimum level of taxation for multinational enterprise groups and large-scale domestic groups in the Union, and in Italy with the application of the global minimum tax from January 2024 pursuant to Legislative Decree no. 209 of 27 December 2023, one of the decrees enacting the tax reform enabling law no. 111 of 9 August 2023.

Very briefly, this involves application of a (minimum) tax, which means that company profits are taxed at an effective minimum rate of 15% regardless of the jurisdiction in which they are generated. In other words, if a company records its profits in a jurisdiction in which they are taxed at a lower rate, the country of origin may apply an additional tax to bring the rate up to the global minimum.

Clearly this is a major innovation, targeting multinationals with global revenues in excess of EUR 750 million.

It is in a certain sense an inevitable tax measure, because the economy has profoundly changed over the last few decades with respect to previous periods in history, and tax systems have proven to be entirely inadequate. You just have to think that the essential framework of the taxation of multinationals dates back to the research carried out by the League of Nations in the 1920s.

Such a tax brings with it technical difficulties that are not to be underestimated.

To mention but a few, the agreement on the common tax rate refers only to the effective tax rate, which will therefore take into account subsidies, deductions and credits, all of which are up to the individual countries, and will require an agreement about definition of the global taxable base, rules governing losses, definition of the types of taxes paid in the country where the relevant income was produced and the tax obligation responsibilities of locally based group enterprises.

The tax measure therefore creates a number of issues around its implementation, not only because it will require full (and difficult) convergence between national tax laws on a global scale, but also because European directive 2022/2523 requires application of the tax also to "non-multinational", and therefore purely domestic, corporate groups.

This is a fact that may undoubtedly also involve entities operating in megaprojects which, once they have reached the revenue levels covered (EUR 750 million), could find they are subject to the global minimum tax.

It could be prejudicial in terms of the tax-related competitiveness of member states, especially with respect to non-EU countries. In fact, in the event that the majority, or in any case some, of the non-EU countries should decide not to apply the global minimum tax also to domestic groups, many European groups, including those involved in megaprojects, could decide to relocate to those countries in order to avoid being subject to the tax.

However, this is not all.

Implementation of the GMT clearly requires a strengthening of the relationship between tax authorities and taxpayers.

The widespread implementation of this tax requires an advanced reporting and monitoring infrastructure for all enterprises, including those working on megaprojects. At the same time, tax authorities are being called upon to apply a totally new and atypical tax. The tax authorities therefore need to operate as consistently as possible with international regulations and international practice, also in order to avoid complex disputes of which, given also the lack of precedent on the interpretation and application of such a tax, the outcome would be entirely unpredictable and therefore uncertain.

Finally, introduction of the global minimum tax will undoubtedly help increase the complexity of managing tax risk. This applies to all enterprises within the scope of the tax and, therefore, theoretically also those operating in megaprojects.

Enterprises these, that are being called upon to equip themselves with new resources and new knowhow in order to offset the uncertainty that introduction of the global minimum tax will inevitably bring with it.

4 The New Controlled Foreign Company (CFC) Rules and Their Significance for Megaprojects

Legislative Decree no. 209 of 27 December 2023, which enacted enabling law no. 111 of 9 August 2023, not only transposed Council Directive 2022/2053 on the global minimum tax, but also updated the Italian regulations governing the Controlled Foreign Company (CFC) regime referred to in article 167 of the TUIR.

Specifically, the decree enacts the provisions of article 3(1)(f) of the Italian tax reform enabling law, which revise the criteria for determining income taxable in Italy.

The purpose is to simplify and rationalise the CFC rule and to coordinate it more effectively with the introduction of the global minimum tax.

The new regulations apply from the tax period following the one in which the decree was approved (i.e. 2024). As explained, they update article 167 of the TUIR, without altering the underlying principle of the CFC regime, which is the criterion of attributing controlled foreign company income according to the principle of transparency.

Therefore, the above attribution criterion will continue to apply in the new legal framework, provided:

1. The controlled foreign company is subject to effective taxation that is less than the "congruent" level set out in article 167(4)(a) of the TUIR.
2. More than one-third of the income generated by the controlled foreign company can be classified as passive income (income mainly of a financial nature) as specified in article 167(4)(b) of the TUIR.

The main innovations in the enabling legislative decree are simplifications.

In particular, Legislative Decree no. 209 of 27 December 2023 facilitates the process of assessing the taxable income of the controlled foreign company by establishing that it is subject to the CFC regime where the effective tax calculated on the basis of the balance sheet figures – i.e. the ratio between the sum of current tax and deferred tax assets/liabilities and the pre-tax profit – is less than 15%.

More specifically, the wording of the decree significantly changes article 167(4)(a) of the TUIR, establishing that the CFC regime applies to controlled foreign companies on two conditions: that they are subject to less than 15% tax and that the financial statements of the controlled foreign company are audited and certified by professionals authorized by the country in which the company is based.

This new criterion (in article 167(4)(a) of the TUIR) replaces the "ordinary" one according to which the controlled foreign company is subject to CFC rules if effective taxation is less than half the virtual domestic tax rate, i.e. the tax the controlled foreign company would have had to pay if resident in Italy.

But this is not all.

Article 3 of Legislative Decree no. 209 of 27 December 2023 introduces a new provision that simplifies calculation of the effective taxation of the controlled foreign company, in relation to both the tax base and the tax rate to calculate. Instead of the ordinary regime set out by article 167(4)(a) of the TUIR, article 167(4-*ter*) offers resident controlled companies the opportunity to pay a substitute tax, instead of income tax, equating to 15% of the net accounting profit for controlled foreign companies, which is calculated without taking into account the taxes that contributed to calculation of the value, the impairment of assets or allocations to the provision for risks.

It is immediately clear that the optional CFC regime – which lasts for three financial years of the controlling entity and is irrevocable – allows the resident entity to avoid both attribution of the controlled foreign company income according to the principle of transparency, and comparison between the effective foreign tax rate and virtual domestic tax rate.

As noted in the literature, the reformed legal framework applicable to controlled foreign companies can rightly be defined as one of the most incisive international tax policies for combating improper tax practices and, together with the global minimum tax rules, takes a further step towards harmonizing national laws and laws to combat profit shifting. While the laws governing controlled foreign companies have the sole objective of avoiding such companies being used to avoid tax, the laws on subjecting corporate income to a global minimum tax aim to prevent the phenomenon. Therefore, these laws operate at a logically earlier level than the CFC laws, which act on cases of avoidance ex-post.

Again, the link between the reformed CFC laws and megaprojects is immediately obvious.

It is perfectly possible for Company A, resident in Italy and involved in the implementation of a megaproject involving other large economic entities operating on a global level and active in a variety of fields (engineering, architecture, industry and finance), to use the services of Company B, which it controls and which is situated abroad.

It is clear that, in such as case, the reformed legal framework would make it simpler to determine the effective taxation of the controlled foreign company, both as regards the tax base and the tax rate to apply.

However, this depends on financial statements being subject to auditing and certification by authorized professionals in the country in which the controlled foreign companies are located, the results of which are used by the controlling company's auditors to issue an opinion on the annual or consolidated financial statements.

5 The New Reshoring Provisions: Tax Leverage as a Tool for Attracting Megaprojects to Italia

Legislative Decree no. 209 of 27 December 2023 has already been discussed in detail. Among the innovations was reshoring, described in article 6.

This is an incentive introduced to encourage the transfer of business operations carried out in non-EU member states or EEA countries to Italy, by offering a 50% reduction on the income subject to income tax and revenue subject to the regional business tax (IRAP) for the tax period in which reshoring takes place and the following five tax periods.

The measure has a clear objective: to promote economic activity in Italy and to reshore multinationals that have relocated their production cycle abroad.

Article 6(1) specifies that the measure applies to economic activities such as business activities and the exercise of arts and professions in an associated form.

Activities carried out in Italy during the twenty-four months prior to their transfer are excluded, and in such cases the incentive does not apply.

Equally, the incentive no longer applies in the event that the activities subject to the incentive are re-transferred (even partially) outside of the EU during the five tax periods following the end of the scheme. In this case, the tax authorities will proceed to recover the tax not paid due to the incentive, together with interest (recapture clause).

If reduction of the tax base takes part during the tax period during which the activity has been transferred and for the following five tax years, the incentive should last for six financial years, if properly planned.

Article 6 of Legislative Decree no. 209 of 27 December 2023 refers to the "transfer of economic activities to Italy", which strictly means the transfer to Italy of an enterprise, or production and commercial activities attributed to companies or divisions, or exercised through the management of an organized complex of assets. The scope of the initiative would not therefore include the transfer to Italy of individual assets not linked to the exercise of an autonomous business enterprise.

The reasoning behind the reshoring incentive regime is clear. It translates to a measure aiming to encourage the transfer to Italy of activities effectively capable of producing market changes through the goods and/or services they offer. This is the reason why activities merely involving the passive holding of assets are not covered by the incentive.

The measure – which still has to be implemented and is currently under examination by the European Commission to determine whether it should be classified as state aid – would undoubtedly also have positive repercussions on the megaprojects sector.

In the event that article 6 of Legislative Decree no. 209 of 27 Decembre 2023 is implemented in full, various companies not currently operating in Italy could well decide to carry out high-impact economic activities such as megaprojects.

This would enable them to enjoy the tax benefits offered by the measure, which would evidently act as a tool to attract entities capable of producing high levels of taxable revenues that are currently resident abroad.

6 The New Cooperative Compliance Regime and Its Implications for Megaprojects

The cooperative compliance regime was first introduced into the Italian legal system with Legislative Decree no. 128 of 5 August 2015 (known as the legal certainty decree).

The decree established the cooperative compliance regime between the tax authorities and taxpayers meeting certain size, structural and organizational requirements. Its purpose was to promote the adoption of stronger forms of communication and cooperation and to facilitate the prevention and resolution of possible disputes.

The scheme was designed for taxpayers with a system for the identification, measurement, management and control of tax risk. In this sense it was (at least originally) presented as a scheme intended for large manufacturing groups, as well as for large banking, financial and insurance groups.

The law which introduced and regulates cooperative compliance established a regime based on the attribution of duties to the tax authorities and taxpayers, with a view to providing certainty about the company's tax risk and ensuring greater transparency in the relationship between the two sides.

A system of monitoring and preventive exchange of information with the tax authorities provides clarity with respect to conduct that could lead taxpayers to operate in breach of tax laws, enabling them to operate within precise margins and to reduce the risk of a possible involuntary breach of tax laws to a minimum.

This alone is sufficient to show the importance of the scheme for entities operating in the megaprojects sector, for which taxation is not a straightforward but a complex issue, involving a number of potential problems (and risks).

This type of entity is naturally called upon to define with the tax authorities the tax risks they could come up against when planning and later implementing projects with a high financial and social impact within the country.

In fact on 26 July 2024 the Council of Ministers approved Legislative Decree no. 108 of 5 August 2024 (published in Official Journal no. 182 of 5 August 2024). This contains supplements and amendments to the cooperative compliance regime as set out by Legislative Decree 128/2015 and most recently amended by Legislative Decree no. 221 of 30 December 2023, which came into force on 4 January 2024.

As well as confirming reduction of the annual turnover threshold for admission to the scheme from EUR 1 billion for 2023 to EUR 750 million for 2024, with further gradual reductions to EUR 500 million from 2026 and EUR 100 million from 2028, Legislative Decree no. 108 of 5 August 2024 made various amendments to way the regime is implemented.

More specifically:

- Entities admitted to the cooperative compliance scheme are no longer subject to the previous exclusion from criminal liability, and instead are subject to the non-application of provisions concerning false tax declarations (pursuant to article 4 of Legislative Decree 74/2000) and breaches of tax laws caused by tax risks that have been the subject of preventive communication (by means of an abbreviated form of consultation or communication).
- Application of these benefits of non-application of the provisions concerning false tax declarations and non-application of the tax penalties are also extended to taxpayers admitted to the cooperative compliance regime which, within 120 days of admission, opt to communicate tax risks relating to conduct during tax periods prior to admission to the scheme, before being informed about any tax inspections. This is a significant change, given that the previous formulation of the law only involved reducing penalties by half in such cases.
- Taxpayers belonging to a group of companies (understood as a group of companies subject to shared control in accordance with article 2359(1)(1 and 2) and article 2359(2) of the Civil Code), where at least one of the companies meets the size requirements for access and provided a certified group tax control framework has been adopted, may access the scheme, eliminating the need for the previous tax consolidation option.
- Entities not meeting the requirement for admission to the regime but that decide to adopt the optional tax control framework for the identification, measurement, management and control of tax risk, and inform the tax authorities of the fact, qualify for

extension of the main benefits for entities opting for cooperative compliance, i.e. the non-application of tax penalties and elimination of the offence of making a false tax declaration.

The condition for access to these benefits is for the tax risks to be communicated by means of a standard consultation, before the presentation of tax returns or tax deadlines, provided the taxpayer's conduct is exactly as set out during the consultation. Such cases also exclude tax breaches involving false or fraudulent conduct or depending on the inclusion of non-existent liabilities in annual financial statements.

Especially when seen in the light of the most recently introduced corrective mechanisms, the cooperative compliance regime has several points of contact with the implementation of megaprojects. Strengthening forms of preventive and transparent communication between the tax system and taxpayers – especially in the case of multinationals involved in major projects – can only lead to positive results on the global market.

After all, when a major company that operates in the international arena, employs large numbers of people and invests capital decides to promote and be actively involved in a large infrastructure, for example, it can only benefit from an advanced compliance system such as the new cooperative compliance scheme. This allows it to be fully aware beforehand of the potential tax risks associated with its activity and of the relevant tax laws.

The need to implement forms of cooperative compliance between companies interested in being involved in megaprojects and the tax authorities is a key necessity also in light of the dissemination within Europe and around the world of mechanisms not unlike that of Italian cooperative compliance (for example, in Australia, Finland, Netherlands, UK and France).

In the reformed legal framework, cooperative compliance is even more of a safeguard for those operating in the area of megaprojects. This is especially true in light of the safeguards and guarantees that it provides for taxpayers, and also for the tax authorities, if the former are admitted to the scheme.

What is more, given the gradual expansion of the scope for access to the regime, there is likely to be a constant increase in interest on the part of economic operators – including those involved in megaprojects – who have been hitherto unable to apply for admission to the regime as they did not meet the revenue requirements of the law. Also in light of these considerations, the reform appears to set out a more definite tax framework that is aiming to attract significant investment into Italy by entities, like those operating in megaprojects, capable of producing high levels of taxable revenues, and above all of creating value in terms of both services and jobs in Italy.

7 Conclusions

The recent reform of the Italian tax system has evidently taken a two-pronged approach. On the one hand it has worked to simplify the existing tax regulations in order to facilitate investment and incentivize the establishment in Italy of entities able to create value for the country. On the other, it has instigated an increasingly decisive implementation of the relationship between tax authorities and tax payers, tax compliance and reduction of corporate tax risk, including by introducing tax laws that ensure greater legal certainty, which will have clear effects on competitivity and the growth of our economy.

It is precisely the most recent amendments to the cooperative compliance regime that are evidence of the movement towards the need for the early definition of high tax risk situations, together with the facilitation of preventive discussions between taxpayers (and no longer only large-size entities) and the tax authorities.

These are the keys to interpreting the tax reform: the prevention of disputes in the relationships between the tax authorities and taxpayers, and the leverage of tax for promotional purposes, and to provide rewards and incentives.

A new style of tax system is taking shape, in which transparency and the good faith of taxpayers, legal certainty surrounding tax law and the use of taxation as an investment incentive can be fully recognised and put to good use, including in the relationship with the tax authorities.

References

1. Abruzzese G, Bonanno G (2023) L'avvento della global minimum tax e i necessari adattamenti dell'ordinamento nazionale. Il fisco 13:1254
2. Allena M (2023) Megaprojects. In: Taxation and the (New) Relationship Between the Italian Revenue Agency and Taxpayers. Springer, pp 203–214
3. Allena M (2024) Il regime di adempimento collaborativo nella recente riforma tributaria. Diritto e pratica tributaria internazionale 2:434
4. Allena M (2024) Quali frontiere per la tassazione delle grandi corporations. Vita e Pensiero 2:66–71
5. Arginelli P, Ronco SM (2024) Cenni di inquadramento sulla riforma della residenza di società ed enti. Diritto e pratica tributaria internazionale 2:548
6. Autuori GA (2023) Lo stato dell'arte della global minimum tax: un panorama internazionale. Riv Trim di diritto tributario
7. Avegno I (2024) Riforma della fiscalità internazionale: la Global Minimum Tax. Amministrazione & Finanza 2:25
8. Avolio D, Piazza M, Volante M (2024) Reshoring delle attività: il punto di Assonime e i possibili riflessi valutativi. Il fisco 12:1099
9. Belluzzo L, Mastrototaro I (2024) La "nuova" residenza fiscale delle società e degli enti nella riforma fiscale italiana. Fiscalità e Comm int:18–29
10. Bomba R, Serena A (2024) Riforma in materia di fiscalità internazionale: la nuova disciplina delle CFC. Amministrazione & Finanza 7:20
11. Buzzi V, Sorci G (2024) Il punto sulla nuova disciplina del reshoring alla luce dei commenti rilasciati da Assonime. In: Rivista di Diritto Tributario Telematico, p 16
12. Canola D, Ludovici P (2024) Italy introduces new linking rules to define tax residence. Int Tax Rev
13. Cochrane RA, Turner JR (1993) Goals and methods matrix: coping with projects complexity. Int J Proj Manag
14. De Broe L, Dannon JR, Chand V (2019) Comments to Public Consultation Document: Global Anti-Base Erosion Proposal ("GloBE"). Pillar Two
15. Della VE (2024) Revisione dei criteri di determinazione della residenza fiscale di società. Il fisco:434–440
16. Demoro S, Michelutti R, Zecca F (2024) Nuovi criteri di residenza fiscale delle società coordinati con la disciplina internazionale. Corriere tributario:326–335
17. D'Ettorre A, Michelutti R, Prampolini R (2024) Il trasferimento in Italia di attività economiche (reshoring). Corriere Tributario 3:205

18. Dodero A, Ferroni B (2024) Adempimento collaborativo a regime con il Decreto "Correttivo". Il fisco 36(23):3323
19. Dodero A, Ferroni B (2024) Adempimento collaborativo potenziato e possibile estensione del sistema di controllo del rischio fiscale alle PMI. Il fisco 16:507
20. Dulcetti A, Longo A (2024) Interazioni tra residenza fiscale delle holding e reshoring di attività economiche. Corriere tributario:474–480
21. Farri F (2022) Il tramonto del binomio "Residenza-Stabile organizzazione" come criterio di collegamento territoriale per i redditi delle imprese multinazionali. Diritto e pratica tributaria internazionale 2:517
22. Ferranti C, Piccinini G (2023) L'interazione tra global minimum tax e regimi CFC. Il fisco 14:1359
23. Ferroni B (2018) Stabile organizzazione: la disciplina nazionale si adegua al BEPS e introduce la "continuativa presenza economica". Il fisco 7:632
24. Garbarino C, Rizzo A (2023) La prospettiva italiana sulla global minimum tax approvata dal Consiglio UE. Fiscalità & Commercio Internazionale 2:5
25. Liburdi D, Sironi M (2024) Adempimento collaborativo: Codice di condotta, gestione del rischio e obblighi di trasparenza. Il fisco 27:2532
26. Liburdi D, Sironi M (2024) L'interpello speciale per i contribuenti in regime di adempimento collaborativo. Il fisco 30:2827
27. Lio M (2023) Il potenziamento dell'adempimento collaborativo: qualche suggerimento per interventi in corsa. In: Rivista Telematica di Diritto Tributario, p 28
28. Lio M (2024) Adempimento collaborativo: riforma in fase di attuazione e Correttivo in rampa di lancio. Il fisco 28:2649
29. Lio M, Macario E (2023) Il Tax Control Framework fattore chiave del nuovo paradigma dei rapporti Fisco-contribuente. Corriere Tributario 5:445
30. Marinello A (2023) Sovranità dello Stato e Global Minimum Tax, Pisa, p 307
31. Melis G (2014) Le interrelazioni tra le nozioni di residenza fiscale e stabile organizzazione: problemi ancora aperti e possibili soluzioni. Diritto e Pratica Tributaria 1:29
32. Melis G (2023) La cooperative compliance: una visione di sistema. Diritto e Pratica Tributaria 2:351
33. Pepe F (2021) L'uso della soft law nel governo delle relazioni fiscali internazionali: funzioni, limiti e possibili prospettive di evoluzione. Rivista di Diritto Tributario Internazionale 2:35
34. Pesiri S (2023) Pillar two e global minimum tax: considerazioni sulla compatibilità della riforma OCSE con il diritto dell'Unione Europea. Diritto e pratica tributaria internazionale 2:519
35. Pistone P (2019) The OECD Public Consultation Document 'Global Anti-Base Erosion (GloBE) Proposal – Pillar Two': An Assessment. Int Tax Law 74(2):63–65
36. Pistone P (2023) La nuova disciplina sulla residenza delle persone fisiche e delle persone giuridiche nel sistema di imposizione reddituale. Diritto e Pratica Tributaria Internazionale 1:872
37. Pistone P (2024) L'europeizzazione del diritto tributario e la riforma fiscale italiana: dall'adattamento all'adeguamento al diritto europeo. Diritto e pratica tributaria internazionale 2:303
38. Savorana A, Vismara F (2024) Prime considerazioni e spunti sulla nuova residenza fiscale delle società (in una prospettiva anche sovranazionale). In: Rivista Telematica di Diritto Tributario. Accessed 8 Apr 2024

The New "*Right of Questioning*" and Its Effects for the Megaprojects Sector

Andrea Purpura(✉)

Università Cattolica del Sacro Cuore, Milan, Italy
andrea.purpura@unicatt.it

Abstract. The recent Italian tax reform reformed the s.c. "*right of questioning*" the Italian Revenue Agency, i.e. the right granted to taxpayers, both individuals and corporations, to ask for clarification to the Italian tax administration on specific and personal cases concerning the cases set out under art. 11 of the Statute of Taxpayers' Rights. If, on the one hand, the reform resizes the "*right of questioning*" available to individuals, on the other hand, it reserves the availability of the so-called "*probatory right of questioning*" to the corporations that adhere to the cooperative compliance regime or that submit a "*new investment right of questioning*". This aspect denotes unequivocally an increasing tendency of the Italian legal system towards the early settlement of tax relations with taxpayers having a high ability to pay. This is clearly a circumstance that also affects those operating in the megaprojects as a sector that has, because of its nature, a very high economic and fiscal importance.

1 The New "*Right of Questioning*" in the Perspective of Megaprojects

The tax reform that has recently affected the Italian tax system has aimed to achieve different objectives: stimulating economic growth, increasing the efficiency and structure of taxes, reducing the tax burden, preventing tax evasion and tax avoidance, as well as implementing a wide-ranging rationalisation and simplification of the tax system, also through the identification and elimination of micro-taxes.

In order to achieve these 'macro-objectives', the reform intervenes across the board in the tax system, reviewing the tax system as a whole.

In order to achieve the ambitious objectives that the reform intends to pursue, it also intervened on the Statute of Taxpayers' Rights.

In more precise terms, as has been more extensively highlighted elsewhere, the Statute has been amended to provide for the strengthening of the obligation to state reasons for tax acts, to further enhance the principle of the taxpayer's legitimate expectations and the principle of legal certainty. There is also a generalised application of the principle of contradictory dialogue between the tax authorities and taxpayers, the

The essay was awarded with the "Best Young Paper" prize by the Scientific Committee of the "MEGAPROJECTS RESEARCH INTERDISCIPLINARY TEAM MERIT CONFERENCE – 2024

E. Favari et al. (Eds.): MeRIT 2024, LNCE 600, pp. 87–96, 2025.
https://doi.org/10.1007/978-3-031-92797-3_7

strengthening of the exercise of the power of self-assessment, as well as the provision - for the first time in the Italian tax system - of the principle of proportionality.

In this perspective, the revision of the Statute of Taxpayers' Rights, which is obviously very wide-ranging, has also passed through a "rationalisation" of the right of questioning.

This institute, in fact, has been profoundly rethought with a focus on reducing its use, especially by 'small taxpayers', while implementing the alternative means of dialogue between the tax authorities and taxpayers also strengthening the prohibition on the submission of appeals, reserving the admissibility of the latter only to those issues that have not been resolved in interpretative documents that have already been issued.

The 'rationalisation' of the right of questioning implemented by Legislative Decree No. 219 of 30/12/2023 passes through a new way of understanding the institution. So much so that, as a general rule, the usability of this will be limited only to cases in which individuals and smaller taxpayers cannot obtain written answers by means of rapid interlocution services, realised also through the use of digital technologies and artificial intelligence.

In addition, and lastly, it is planned to make the admissibility of interpellation requests subject to the payment of a fee, to be graduated in relation to various factors, such as the type of taxpayer, the value of the matter being interpellated and the volume of business or revenue achieved by the taxpayer.

But there's more.

If it is true that the right to question the Tax Administration currently regulated by art. 11 of the Statute of Taxpayers' Rights has been clearly downgraded with regard to individuals and small taxpayers, different conclusions are drawn with respect to large taxpayers.

And thus also with regard to those taxpayers who usually operate in the megaprojects sector, which, as is well known, make very large investments and produce significant effects both in terms of employment and on the environment where they are carried out.

If for no other reason than that, among the various types of petitions that may be filed pursuant to Art. 11 of the Statute, the s.c. "probatory"question - a very important type of request for advance knowledge of the applicability or non-applicability of a certain tax regime - has been circumscribed precisely to very large taxpayers.

In this perspective, as will be discussed below, the tax reform further enhances the early dialogue between large taxpayers - and thus also those usually operating in the megaprojects sector - and the Italian tax administration, while sacrificing the anticipated dialog with small taxpayers.

In this sense, in the following pages, particular attention will be paid to the pro-files of the reform of the "probatory" question and, in conclusion, to the s.c. "question on new investments", a request that, although not changed by the recent reform of the Italian tax system, has acquired even greater importance precisely in the light of the revision of Article 11 of the Statute of Taxpayers' Rights.

2 The "*Probatory Right of Questioning*" as Originally Formulated

The category of probatory questions includes all those requests aimed at obtaining an opinion from the tax authorities:

- on whether the conditions for access to a certain tax regime are fulfilled;
- or on the suitability of the evidence required by the rule for access to a particular tax regime.

Article 11 of the Statute of Taxpayers' Rights, before being subject to the 2024 reform of the Italian tax system, pointed in this direction.

According to the provision, the possibilities of filing an evidentiary appeal are limited to 'cases expressly provided for'.

In this sense, the legislator, opting for an approach based on the taxability of the cases, has clarified that the probatory right of questioning, unlike the ordinary one, cannot be activated in relation to any case in which the taxpayer considers an assessment by the tax administration to be useful. On the contrary, the "probatory" request may be activated just in those cases in which such an option is expressly provided for by Article 11 (1)(b) of the Taxpayer's Statute.

Moreover, from a technical point of view, the submission of an application for an evidentiary appeal presupposes that the taxpayer has provided every useful element of assessment in order to obtain the most complete and informed answer from the tax administration.

So much so that, in the presence of petitions accompanied by insufficient evidence to convince the administration of the suitability of the evidentiary framework represented, the Revenue Agency will not be required to send a request for additional documentation to seek new and different evidence in favour of the taxpayer.

This expression must be interpreted in a wide sense.

Therefore, as pointed out in the literature, it must be considered to include not only the hypotheses of access in the strict sense (e.g., applications for access to the worldwide consolidation regime), but also cases in which the non-applicability of certain limitations or the applicability of special rules with respect to those ordinarily applicable is at issue (e.g., the interpellations relating to C.F.C. for which the object of the application is the non-applicability of taxation by transparency).

In this perspective, it is worth recalling how, with reference, therefore, to the petitum of the request, the interpretative questions that could be put forward through the probatory right of questioning concerned multiple and diverse cases, namely:

- requests submitted under Art. 113 of the TUIR by credit institutions concerning the non-application of the participation exemption regime under Art. 87 of the TUIR with reference to participations acquired in the context of interventions aimed at recovering credits deriving from the acquisition of participations or the conversion into newly issued shares of credits towards companies in temporary financial difficulty, in compliance with the supervisory provisions for banks issued by the Bank of Italy;
- the requests for the continuance of the consolidated tax regime, pursuant to Article 124, paragraph 5, of the TUIR, filed on the occasion of certain reorganization transactions otherwise interrupting the regime, aimed at ascertaining that, even after the

carrying out of such transactions, all the requirements provided for in Art. 117 et seq. Remain in force for the purpose of gaining access to the regime;

- requests for the access to the worldwide consolidation (Article 132 (3) of the Consolidated Income Tax Act) in order to verify the existence of the requirements for the valid exercise of the option;
- the applications submitted by 'non-operating' companies as well as the applications of companies with systematic losses (Article 2, paragraph 36-decies, Law Decree no. 138/2011) in order to acquire the administration's opinion on the presence of objective conditions that make the discipline provided for by Art. 30 of Law no. 724 of 23 December 1994 not applicable;
- the applications envisaged for the purposes of recognizing the ACE benefit referred to in Article 1 of Decree-Law no. 201 of 6 December 2011, converted into Law no. 214 of 22 December 2011, in cases of transactions that could potentially result in the undue duplication of benefits (Article 10 of Ministerial Decree of 14 March 2012);
- the requests required for new residents in order to recognize the option for the substitute tax on income produced abroad by individuals who transfer their tax residence to Italy (by Article 24-bis of the TUIR).

It is clear, therefore, that in the former formulation of Art. 11, the 'probatory' request was a non-circumscribed request from a subjective point of view. By contrast, it was - at least until now - a request circumscribed only in terms of its object, because it concerned requests aimed at obtaining clarifications concerning the adoption of specific tax regimes.

Today, as will be highlighted in the following paragraph, the reform of the tax system has lead to restricting the field of application of the probatory request.

As will be said, at a first assessment the probatory question would appear to have been distorted by the tax reform. However, this is not the case, as the new wording of the Institute is coherent with the tax reform as a whole and, more precisely, with the intention of the Italian legislator to encourage the use of the cooperative compliance regime and, in more general terms, to stimulate investments within the State.

2.1 The New "*Probatory Right of Questioning*" and Its Effects on *Megaprojects*

As anticipated above, Legislative Decree No. 219 of 2023 has restricted the probatory question institute to persons in cooperative compliance or who file interpellation petitions for new investments due to the fact that, as indicated in the report to Law No. 111/2023, such persons have a high level of compliance and transparency in the management of tax risks.

It should be clarified immediately that the Italian Revenue Agency will have to clarify the operational implications of this provision with reference to interpellations aimed at obtaining a view on the existence of conditions and the assessment of the suitability of the evidence required by law for the adoption of a certain tax regime.

The decision to limit the applicability of this questionnaire to taxpayers who adhere to the cooperative compliance regime or who file requests for an answer relating to new investments governed by Article 2 of Legislative Decree No. 147 of 14 September 2015 is attributable to different reasons.

As highlighted in the literature, on the one hand it meets the need to deflate the use of the institute. This was already clearly indicated in the Delegated Law for Tax Reform No. 111 of 9 August 2023.

On the other hand, the consideration that probatory requests, perhaps even more so than disapplicative ones, require the assessment of a series of factual elements that are difficult to carry out in an advisory capacity. On the other hand, the decision to retain the possibility of submitting evidentiary appeals only in a subjectively circumscribed area seems to be justified by the high level of compliance that characterises taxpayers who adhere to the cooperative compliance regime or who make new investments available to taxpayers.

Reasoning in this way, the new wording of Article 11 of the Statute of Taxpayers' Rights seems to deviate - at least, at first glance - clearly from the previous wording of the provision that has, to date, regulated the probatory question.

More precisely, the probatory interpellation would seem to have been completely re-designated by the reform of Art. 11. and nothing else, because recourse to the institute would be limited.

On the one hand, consistent with the 'dimensional' limits provided for access to the collaborative compliance regime, to resident and non-resident persons (with a branch office in Italy), who realise a turnover or revenue of no less than EUR 1 billion.

On the other hand, to investors, Italian or foreign, who intend to make major investments in the territory of the State, with a value of no less than EUR 20 million, with significant and lasting repercussions on employment.

If this is true, the envisaged narrowing of the perimeter of applicability of the probatory right of questioning appears to be only apparent and consistent with the animus of the proxy for tax reform.

The one referred to does not, in fact, represent a new case which may be the object of the petition. On the contrary, it is a petition that was already submitted to the tax authorities under the previous wording of Article 11 of the Statute of Taxpayers' Rights.

However, the identification of the aforementioned hypothesis as a case that may be the subject of an interpellation must undoubtedly be read in the light of the broader renewal of the subjective perimeter on which the evidentiary interpellation is based.

An institute whose activation will henceforth be linked to (and subordinate to) the establishment of a dialogue with the tax authorities.

In accordance with the foregoing considerations, it is clear that the reform, where it revises the subjective criterion for admission to the probatory appeal, tends to produce two effects.

On the one hand, it favours large taxpayers and therefore, among all, also those operating in the megaprojects sector, which today, even more than in the past, will have an 'exclusive' canal - because exclusively reserved to them - to dialogue with the Tax Administration and define in advance the tax regime applicable to them.

In other words, the legislator has not changed the probatory question from a substantive point of view.

It remains, in fact, a 'closed case' type of request, within the scope of which the legislature continues to refer, even today, to various types of compulsory requests already known to the system.

Nor are the technical features of the request changed, which will therefore continue to take the form of a demand to the tax authorities aimed at obtaining an opinion on the existence of the conditions or the suitability of the evidence represented by the taxpayer in the request for the purposes of access to a certain tax regime.

On the other hand, there is a change in the subjects who will be able to turn to the tax administration; subjects - such as those operating in the megaprojects sector - who, in order to establish a participatory relationship with the Revenue Agency and to have greater certainty on their operations in Italy and on the tax treatment reserved for them, will necessarily have to adhere to the regime of collaborative compliance or submit an interpellation on new investments.

An important change, therefore, which serves as an incentive to enter the cooperative compliance regime and which must be read in conjunction with the planned reduction of the thresholds for access to the regime.

And in fact, the regime, which until now was reserved for the largest taxpayers, before the changes envisaged by the reform, for the years 2022, 2023 and 2024, was reserved for resident and non-resident taxpayers with a turnover or revenue of no less than EUR 1 billion.

Article 7, paragraph 1-bis of the '*Cooperative Compliance Decree*', as reformulated by the '*Cooperative Decree*', now admits taxpayers with a turnover or revenue volume to the regime

- from 2024, not less than EUR 750 million
- from 2026, not less than €500 million;
- from 2028, not less than €100 million.

Pursuant to Paragraph 1-ter below, the size requirements set forth in Paragraph 1-bis are assessed by taking, as a reference parameter, the higher of the revenues indicated, in accordance with proper accounting principles, in the financial statements for the financial year preceding the one in progress on the date of submission of the application and the two preceding financial years and the turnover indicated in the value-added tax return for the previous calendar year and the two preceding calendar years.

In this perspective, it is clear that where the reform has revised the subjective perimeter of the evidentiary interpellation, it has undoubtedly aimed to achieve very precise objectives, namely:

- enhancing the preventive check on the tax regularity of the activities carried out by major taxpayers exclusively;
- strengthen the dialogue between the latter and the tax authorities, including through the establishment of a dialogue on the entrepreneurial choices and operations of companies participating in the scheme, in order to reach a common assessment of situations that may generate tax risks.

In this sense, the recent reform of the probatory petition - unlike what the legislator has done with other aspects of the petition, which, on the contrary, raise perplexity but which, for thematic reasons, we will not go into here - would not seem to give rise to particular concerns.

On the contrary, it goes precisely in the direction desired by the legislator and towards which the system seems to be leaning more and more: a 'functional' tax, aimed at placing the contribution to public expenditure (pursuant to Article 53 of the Constitution) in a perspective of orderliness, which allows to lower the tones of conflict and contrasting interests between the individual and the State.

In other words, the revision of the probatory question in terms that are clearly favourable to large economic operators would seem to testify to the ever-increasing urgency of change in the approach to the relationship between the taxpayer and the tax-man, whereby the simplification and certainty of the aforesaid relationship is becoming the rule and no longer the exception.

3 The Renewed Centrality of the "*New Investment Right of Questioning*"

It has been highlighted within the literature that the recent tax reform of the Italian tax system aims at overcoming all those aspects that separately and together contribute to disincentivising investment and increasing operating costs.

In other words, a tax reform that intends to go beyond the traditional reluctance of a 'cautious' and 'mistrustful' financial administration towards taxpayers and, at the same time, a tax litigation that is no longer the main forum for the ordinary discussion between financial administration and taxpayers.

In this perspective, it may be said that the reform is also placing again at the centre of the 'tax-taxpayer' relationship another type of question, i.e. the one concerning new investments.

This query is of particular relevance for entities operating in the megaprojects sector, since it is a flexible instrument of pre-trial dialogue with the tax authorities, aimed at establishing the various tax aspects of an investment initiative being planned.

It is a query that, perhaps more than any of the other types provided by art. 11 of the Statute, implements precisely what the reform calls for: taxation not as an instrument of uncertainty, but as a driving force for encouraging investment in Italy.

The centrality of this type of question emerges in some fairly recent data: suffice it to say that, according to reports from the Internal Revenue Service, in 2022 alone this institution was activated in relation to 19 new entrepreneurial initiatives, with the prospective creation of more than 2500 jobs.

It is not, therefore, just a question of clarification.

Rather, it is a genuine business planning tool capable of significant effects, and benefits, in our country.

This aspect of the new investment question clearly emerges in the forms and purpose of the petition, which may be proposed by domestic and foreign companies and may concern:

- the tax treatment of their investment plan;
- the tax treatment of any extraordinary transactions that are hypothesized for the realization of such plan;
- the assessment as to whether or not a company exists;
- the prior assessment as to whether there is no abuse of fi-scale law or avoidance;
- the existence of conditions for the disapplication of anti tax avoidance provisions and access to any regimes or institutions provided by the tax system.

Regarding the investment, the main requirements seem to be the conse- quences it must produce and the size of it, as it could then materialize in any form.

With reference to the nature of the investment, it has been specified that (Ministerial Decree April 29, 2016) it may concern:

- either the establishment of new economic assets or, alternatively, the expansion of pre-existing economic assets;
- the diversification of the production of an existing production unit;
- the restructuring of an existing economic activity in order to enable the enterprise to overcome or prevent a crisis situation;
- transactions involving holdings in an enterprise.

In order to determine the value of the investment, all financial resources, including those of third parties, necessary for the company to implement the investment plan must be considered. Finally, from a formal point of view, as specified by the Italian Financial Administration, the request must be submitted in Italian.

It is clear, then, that the new investment right of questioning fulfills a central role for those who intend to start an activity with significant economic and employment implications in our country.

A circumstance that undoubtedly closely affects precisely those who operate in the set-tore of megaprojects, subjects that undoubtedly have an interest in investing in a framework of legal and tax certainty.

Certainty that the right of questioning on new investments is undoubtedly able to guarantee as long as the - minimal and essential - content of the petitions is respected.

3.1 The Content of the "*New Investment Right of Questioning*"

As noted in the literature, the certainty that typically follows the new investments question is implemented only if the correct content of the petition is respected. The latter, in fact, has a content that we could define as "typical," since it must unambiguously indicate:

- the identifying elements of the enterprise (name, legal representative, registered office or tax domicile, if different from the registered office, tax code or VAT number or other identification code of the enterprise, as well as the indication of the contact details, including telematics, of the domiciliary for the interpellation procedure at which it is requested to forward comunications pertaining to the procedure).In case of more than one participant in the investment, the application must contain the identifying elements of all of them.
- the detailed description of the investment plan, on which the assessment of the Revenue Agency is requested with reference to the tax treatment of the same and the corporate operations planned for its implementation.

The description must necessarily specify, then: the amount of the investment, not less than 30 million euros, and the methodology followed for its quantification; the timing and manner of its implementation; the significant employment effects, in terms of increasing or maintaining the level of employment, to be assessed in relation to the activity in which the investment takes place and lasting; the effects, also in quantitative terms, of the investment object of the petition on the Italian tax system.

On this point, it must be highlighted that the petitioner must indicate exclusively the effects on the Italian tax system that are directly and immediately deriving from the specific plan to which the petition refers, not assuming importance also of the additional and more general effects deriving 'in cascade' from the investment on the Italian tax system and economy (Agenzia delle Entrate Circular 1.6.2016 no. 25, § 4).

Finally, the application must state:

- the specific tax provisions whose interpretations are sought or in relation to which it is asked to evaluate the possible abusive nature of the transactions related to the investment plan;
- the specific anti-avoidance provisions of which disapplication is requested and the specific regimes or institutions to which access is requested;
- the exposition, in a clearly and unambiguous manner, of the tax treatment that the taxpayer considers appropriate with respect to the investment plan, with explication of the solutions and behaviors that the petitioner intends to adopt in relation to its implementation;
- the subscription of the applicant or its legal agent or general or special pro- curator (in which case the power of attorney should also be attached).

In light of the above considerations, it is clear that the new investment petition therefore fulfills an essential role for all those who intend either to enter for the first time into a relationship with the Italian legal system or to corroborate an activity that already exists and has been carried outthere.

A tool of certainty that business actors operating in the mega-projects sector must consider now more than ever, especially in consideration of the probatory question reform, which is precisely reserved for those who have filed a right of questioning on new investments (as well as for those who have adhered to the cooperative compliance regime).

4 Conclusions

The tax reform that has recently affected the Italian tax system incidences in a important way in the relationship between the Revenue Agency and the taxpayers, with the aim of making it, if not more friendly, certainly even more cooperative and mutually reliable, regardless of whether taxes are raised or lowered in terms of higher or lower applicable rates.

The literature reaches these, first, conclusions because of the principles and new guiding criteria regarding the assessment procedure which are destined to revolutionize cooperative tax compliance as previously known, including in the relationship between those operating in the megaprojects set-tore and the tax authorities.

Further evidence of the cultural, even before the legal, change taking place can be found precisely in the reform of the institute of the right of questioning, with specific emphasis on the probatory question.

The latter, as underlined, will henceforth be available only to those who adhere to the collaborative compliance regime or to those who file a right of questioning on new investments.

References

1. Allena M (2024) Il regime di adempimento collaborativo nella recente riforma tributaria. Diritto e Pratica Tributaria Internazionale 2:434
2. Allena M (2023) Megaproject, taxation and the (New) relationship between the Italian Revenue Agency and Taxpayers. In: Cantoni F, Favari E (eds) Complexity and Sustainability in Megaprojects, p 203
3. Barbiani E (2024) L'adempimento collaborativo a confronto con le best practice in ambito di Tax Control Framework. Corriere Tributario 3:323
4. Conte D (2022) Interpello tributario e diritto ad una buona amministrazione: riflessioni per un bilancio. Diritto e Pratica Tributaria 6:2024
5. Dodero A, Ferroni B (2024) Adempimento collaborativo potenziato e possibile estensione del sistema di controllo del rischio fiscale alle PMI. Il Fisco 6:507
6. Dodero A (2024) Il nuovo interpello e i supporti interpretativi che favoriscono il rapporto Fisco-contribuente. Il Fisco 3:231
7. Fortunato L (2024) Riorganizzazioni extra europee al bivio PEX. Il Fisco 29:2773
8. Giovannini A (2023) Notarelle sulle deleghe in bianco e Costituzione. In: Rivista Telematica di Diritto Tributario del 18 maggio
9. Lio M, Macario E (2023) Il Tax Control Framework fattore chiave del nuovo paradigma dei rapporti Fisco-contribuente. Il Fisco 5:445
10. Marino G (2024) L'adempimento collaborativo. In: Melis G, Manzon E (eds) Il diritto tributario nella stagione delle riforme, Pisa, p 185
11. Mastroiacovo V (2023) Procedimenti accertativi e nuovo rapporto tra fisco e contribuente nella legge delega di riforma tributaria. In: Rassegna Tributaria, p 480
12. Melis G (2018) Lezioni di diritto tributario, Torino, pp 399–400
13. Melis G (2023) La cooperative compliance: una visione di sistema. Diritto e Pratica Tributaria 1:351
14. Melis G (2023) Una visione d'insieme delle modifiche allo Statuto dei diritti del contribuente: i principi generali. Il Fisco:4441
15. Patimo R (2024) Disciplina degli interpelli, in One Fiscale, Settembre
16. Purpura A (2024) Alcune riflessioni sul riformato interpello tributario. Diritto e Pratica Tributaria 4:1208
17. Vanz G (2017) Investitori esteri e interpello nuovi investimenti. Rass Tribotest:969
18. Viotto A (2017) Tutela dell'affidamento, consulenza giuridica e interpello. Riv dir trib:690

Negotiating Megaprojects and Good Faith

Francesco Zecchin(✉)

Università Cattolica del Sacro Cuore (sede di Piacenza-Cremona), Via Emiliana Parmense, 84, 29122 Piacenza, PC, Italy
francesco.zecchin@unicatt.it

Abstract. The paper deals with the issue of negotiations in the context of megaprojects with reference to the importance of the good faith principle. After an introduction to the meaning of such a principle, the author attempts to identify both its content and consequences that may result from behaviour in bad faith in the case of negotiations functional to megaproject contracts. Firstly, it is discussed the scenario where good faith imposes obligations which do not necessarily relate to negotiations, but are only incidental to them, as might be about knowledge of confidential information. Secondly, it examines cases where good faith imposes obligations that are directly related to negotiations. The most relevant and problematic of these, even in megaprojects, is breaking off negotiations. A distinction is made between situations where the client is a private party and where it is a public authority. Finally, it considers the case where public administration is not the megaproject's client, but it is involved in obtaining the authorisations necessary for works to start. Public administration's conduct might give rise, indeed, to a legitimate expectation of such authorisation, which is ultimately not granted, even if lawfully at the level of administrative discretion, but which causes harm to private party.

Keywords: Megaprojects · Negotiations · Good Faith · Precontractual Liability · Pre-action Liability · Private Client · Public Client

1 Megaprojects and Good Faith

The topics of good faith and precontractual liability are a classic of legal studies on contracts (Castronovo, 2018), and they have been the focus of extensive debate in recent years, which has been followed by some regulatory and case-law developments, too (Mantovani, 2011). It might therefore be thought that new rules and the discussion that have taken place, both in general and in particular, would be sufficient to address these issues also in the specific case of megaprojects contracts. On the one hand, however, there is already a need for specifications regarding the different types of contracts, since very dissimilar outlines are clear. On the other hand, megaprojects "are not just magnified versions of smaller projects", but "a completely different breed of project in terms of their level of aspiration, lead times, complexity, and stakeholders" (Flyvbjerg, 2014). It is enough to consider how long, and complex are negotiations that lead to conclusion of contracts by which megaprojects will then be implemented, compared to consumer

E. Favari et al. (Eds.): MeRIT 2024, LNCE 600, pp. 97–107, 2025.
https://doi.org/10.1007/978-3-031-92797-3_8

contracts and the normality of business-to-business contracts. This gives rise to the need, or at least the opportunity, to explore the subject in a specific key.

From this perspective, the first question that we want to address is: what does it mean to behave in good faith during negotiations, and in negotiations for megaprojects contracts, particularly? Secondly: what happens if you do not act in good faith?

Well, as far as good faith in general is concerned, such term comes to mind: fair play, loyal person, the *honeste vivere* of Roman times. But these are already very vague concepts, and even more so in our modern society where traditional models of behaviour are being strongly questioned (Zecchin, 2020). A process of the characterisation of good faith is even more necessary today. It is said, not by chance, that good faith is a general clause, if not the most important, then certainly one of the most important general clauses. And this means precisely that its value as a legal rule must be determined case-by-case basis according to the actual situation (Mengoni, 2011a). In other words, the legal case must be completed in the light of the real case, but at the same time, to assess whether behaviour is contrary good faith, it is not possible to rely directly on the prevailing conduct in society, but to rely on the value standard that good faith recalls, also in the light of case-law (Nicolussi, 2021).

Namely, on the one hand, the meaning of good faith is revealed when it is applied to a real situation, on the other hand, this does not mean homologating the way the community normally behaves, which could well be completely contrary to good faith. As it has been said, good faith limits autonomy of parties not in authoritarian terms, but by screening social values within the legal framework (Castronovo, 1986), i.e. following the logic of universalisation and not that of application to the specific case (Mengoni, 2011a).

Precisely from this point of view, it must be said that in the debate on megaprojects the issue of good faith is not completely ignored, but it is essentially addressed with reference to the execution of contracts. For instance, both the NEC and the FIDIC models - which are fully recognised by the World Bank (Cardinale, 2001) and the European Commission (Besaiso et al., 2018) as the best models for negotiations related to megaproject, and are used by many states, such as the United Kingdom, China and South Africa, for the construction of large public works (Balestra, 2016) - share an increased emphasis on "mutual trust and cooperation" between contract's parties (Close, 2017).

However, while the NEC contracts explicitly propose the adoption of a clause introducing good faith as general rule in relations between parties (10.2), there is no such a solution in the FIDIC models. Of course, this does not mean, of course, that parties choosing one of the latter models are encouraged to behave improperly: many FIDIC clauses are based on good faith. For instance, according to the Clause 18.2, the contractor must give notice, in the proper form, within 14 days of becoming aware, or of the date when it should have become aware, of the exceptional event which makes its performance impossible or leads to a delay.

Outside of such cases, i.e. where the specific contract does not contain a clause that expressly providing for good faith, there is a high risk that that the victim of wrongful conduct will be left without protection, which will vary according to the law applicable to the contract. In other words, depending on the law applicable to the contract, the conduct may or may not be a source of liability.

In England and Wales, for example, case law has repeatedly made clear that there is no general doctrine of good faith in contract law (Glover, 2017). This means that in the matter of the misleading reliance on an exculpatory clause, even where such reliance is implied in the contract, the judge normally has no general tool to prevent the other party's improper behaviour, although such a result can sometimes be achieved by other legal means (Castronovo, 2001).

The situation is far apart when the applicable law is one of a legal system which recognizes the principle of good faith among the cardinal ones managing contracts. Among these, German law stands out, but there is the Italian one, too: the Article 1375 of the Italian Civil Code ruling in general that the parties of a contract must behave properly during its performance. So, in the example just given, the misleading reliance of liability exemption clause would probably be considered unlawful. As it would be in the case of a request to terminate a contract based on a minor breach (Albanese, 2021). And this is the case even if the parties, in somewhat enigmatic terms, had expressly established that the principle of good faith does not apply between them, since such an exception cannot be made in the Italian system (Piraino, 2015).

It is important to remember, however, that the importance of good faith arises long before the conclusion of a contract. In such cases the words "good faith in negotiations" are used. Indeed, some legal systems expressly recognise the necessity for the parties to behave fairly when negotiating a contract. This is particularly the case in Italian law, which expressly provides that the principle of good faith also applies in the negotiation phase. To be more detailed, as in other European jurisdictions, the Italian Civil Code provided that "the parties must act in good faith in negotiating and concluding contract" (Article 1337).

Here, too, good faith has a non-specific meaning, which may be understood as the obligation to behave in such a way as not to damage other party in negotiations (Albanese, 2018). Nevertheless, an attempt can be made to define its meaning more precisely by identifying typical situations.

A first set of cases that can be outlined is that of the obligation to prevent harm to the other party during negotiations at a level that is wider than conclusion or content of a contract. In other words, these are cases where the breach of good faith may be linked to negotiations (Benatti, 1963) but may also have been simply caused by them. So much so that it has been suggested, it has been suggested not to speak of precontractual liability (Roppo, 2011). However such a classification, if understood to be not tied only to functional obligations for stipulation or determination of contract's content, could still be continued, since it is always a matter of behaving in good faith during negotiations (Albanese, 2018). After all, the Article 1337 of the Italian Civil Code binds to act in good faith both in the 'negotiations' and in the 'formation of the contract'.

Perhaps the most striking example is the obligation of non-disclosure with regard to confidential information about the other party, that has become known as a result of negotiations (Gallo, 2004). This may be the case either with regard to confidential information that is functional to bargaining or specification of the contract's content, or with reference to confidential information that has nothing to do whatsoever with negotiation or agreements, but which has occasionally come to light during negotiations. For the former, details of megaproject that must not be disclosed, for the latter, names both of people involved for private or public companies in the implementation of the megaproject and identity of other possible players. In any case, the obligation of good faith means that such information cannot be disclosed or used for any purpose other than contract's stipulation.

In case of megaprojects, actually, confidentiality agreements are often signed in advance to address this issue. To the truth, in legal systems that impose an obligation to behave fairly, and to some extent also in common law (Allen, 1990), there would be no need for such agreements in the abstract, since obligation of confidentiality would be clear from legal framework or case law (Castronovo, 2001). However, from a practical point of view, confidentiality agreements are extremely useful - as has been demonstrated by common practice even in systems where they are expressly provided for, such as the French system following the recent reform of the Code - because they provide an accuracy that the law is not always able to provide (Zimatore, 2010). From this perspective, the main issues that come to the fore are duration, the content of what must not be disclosed and compensation for damages (Federico, 2018).

With regard to duration, the law is not usually explicit. However, non-disclosure agreements must do so. Otherwise, they may not be invalid, but only enforceable for the period that may have been established, directly or indirectly, in the applicable law at the time of negotiations (Confortini, 2017).

As far as non-disclosure agreement's content is concerned, in the field of megaprojects, confidential information that could become known regards, for instance, details of works, or identity of clients and players involved. Once again, therefore, an *ad hoc* agreement would provide a level of detail that the law cannot. And where it would be too general from this point of view, it would run the risk of being invalid due to the undefined content of the confidentiality agreement (Roppo, 2011).

The advantage of a confidentiality agreement in place of the law on compensation for damages is self-evident (Federico, 2018), since the entity of the damage is established in advance by the parties through a penalty, which also avoids the need to prove the causal link between the damage and the disclosure. It will of course be necessary to identify an extent of the penalty that is not, to use the words of the Italian civil code, 'clearly exaggerated' (Nicolussi, 2021) or, to use the common law terminology (Halson, 2018), a penalty clause (invalid), but instead constitutes liquidated damages (valid). As a result, the aggrieved party only has to plead the breach of the agreement, and this will lead to the right to be paid the amount provided for in the agreement. Since it is an obligation, the penalty will not be due if the other party succeeds in proving that the disclosure was due to a reason not dependent on it or, in certain jurisdictions, due to force majeure.

At least in Italy, even in the absence of a confidentiality agreement, the rule of liability would still be that of non-performance and not that of tort, since an obligation, namely, to act in good faith, has been breached (Castronovo, 2001). As a consequence, the party whose confidential information has been disclosed might only have to prove the damage suffered and the connection with the disclosure of such a confidential information (causal link). It would then always be for the other party to prove impossibility without fault or force majeure.

Again, in the case of non-agreement, there may be cases where the other party has benefited from the disclosure. This point, which in the case of a penalty would instead be included in the amount agreed by parties, would be ruled by the law of unjust enrichment applicable during negotiations (Castronovo, 2007).

2 Precontractual Liability and Breaking Off Negotiations in Private Megaprojects

However, good faith in negotiations is also, and especially, a matter of obligations that are closely linked to conclusion of contract or its content: their breach gives rise to what is technically known as precontractual liability. In other words, these are breaches of good faith which have a direct impact on negotiations, thereby influencing them and determining the terms of the contract.

Typical situations have also been identified for this case, but before going into the most common one in the megaprojects' context, it should be pointed out that in this such a context the breach of good faith does not lead to the invalidity of the contract, unless in specific cases it is enriched by other elements in addition to the unfairness (Albanese, 2003). Rather, it gives rise to an obligation to pay damages and does not render the contract unlawful, which would lead to nullity or voidability. In short, the rules of contractual liability and not the rules of illegal acts arise from good faith (Mengoni, 2011a, Albanese, 2018).

However, invalidity of contract is an infeasible result with regard to the cases of pre-contractual liability that seem to be the most important in the megaprojects' framework, i.e. breaking off negotiations when one of the parties is actually convinced that contract will be concluded. In other words, there is a party that expects a contract to be concluded, but this does not happen because the other party withdraws from negotiations.

This is a case that is as common as it is problematic, even in megaprojects, because it is necessary to verify whether the party that withdraws its willingness to conclude the contract has behaved improperly, leading counterparty to expect a scenario opposite to what happened (Albanese, 2018). If this were the case, it would be interpreted by Italian law and other civil law jurisdictions, such as German or French, as a breach of good faith, and the party deemed responsible would be ordered to pay damages (Castronovo, 2018).

The key question is when it can be said that a legitimate expectation of concluding the contract has been created, and breaking off negotiations is therefore wrongful meaning that it gives rise to an obligation to pay damages (Castronovo, 2007). In other words, the question is when a party can legitimately believe that the contract has been concluded, but it has not.

The first condition - from an objective point of view - is that negotiations reach a sufficiently advanced stage (Mantovani, 2011) or, in the first phase, on assurances that contract will be concluded by the party that then breaks off negotiations (Roppo, 2011). In short, negotiations must be serious. Indicators of an advanced negotiations, in addition to time, are the exchange of papers, confidential information, minutes or punctuation (Roppo, 2011), while assurances could be documented or in the light of several contacts that have taken place in a short period of time.

The second condition - subjective in nature - necessary for precontractual liability concerns behaviour of the party that has broken off negotiations (Mengoni, 2011b).

Indeed, such a conduct must be contrary to the good faith, so that the breaking off negotiations would not lead to liability if the expectation of the other party was unfounded, i.e. if there was nothing to justify the belief that contract would be concluded. If, on the other hand, one of the parties acts in such a way as to give rise to a legitimate expectation that the contract will be concluded, failure of negotiations means liability. This is all the more so if such a behaviour was already aimed not at the stipulating the contract but at gaining an advantage. For example, it is a breach of good faith if a person enters into negotiations not with the intention of concluding a contract, but only in order to have a better chance of success in another simultaneous negotiation, which is the only one that really counts (Castronovo, 2001).

With regard to liability rules, in most legal systems tort law serves as a reference framework (Castronovo, 2007), but in Italy there is now a growing acceptance (Albanese 2018) that liability must be assessed in terms of breach of duty rules (Mengoni, 2011b), since there is an obligation to behave in good faith. This outcome has practical consequences, e.g. in terms of the limitation period and the burden of proof, which is lower for the party whose confidence has been breached (Mengoni, 2011b), while the party that has interrupted negotiations is not liable if the breach of good faith results from a supervening impossibility of compliance, such as an extraordinary and unforeseeable change in the reference market (Roppo, 2011).

In terms of damages, of course, the issue is not what would have been avoided or obtained by concluding the contract, i.e. the expectation interest, but rather the reliance interest, i.e. the loss recorded in the assets of the party that suffered the failure of the negotiations, both as a loss suffered and as a loss of profit (Castronovo, 2007). In the first case it is necessary to consider the expenses - which in megaprojects are usually considerable - paid for negotiations that turned out to be in vain. In the second case it is necessary to consider the profit margins from lost of business opportunities - which in megaprojects are not so numerous - because they were persuaded to conclude a contract that did not take place. Of course, it will be up to the injured party to prove the amount of such damages and their connection with the other party's breach of good faith.

3 Breaking off Negotiations and Precontractual or Pre-action Liability in Public Megaprojects

However, the area of megaprojects where the issue of breach of good faith in the form of a bad faith breaking off negotiations arises most frequently and in the most specific terms is probably that of public megaprojects. This is illustrated internationally by one of the most famous examples involving a Dutch company whose proposal for the construction of a municipal swimming pool was the best, but an alternative plan was subsequently accepted by the city council on the initiative of a councilor. As a result, the Dutch Supreme Court awarded the company damages for expectation interest (Hondius, 1990).

Indeed, on the one hand, the growth of compensation as a remedy in the last century has also involved public bodies, at one of the crucial points where civil and administrative law share common ground (Nicolussi-Zecchin, 2021). On the other hand, it has become increasingly clear that the inequality between private parties and the public administration, even if the latter is the holder of a discretionary power, must not lead to arbitrary behaviour that justifies conduct contrary to good faith. After all, it is also and above all a crucial obligation of a party that has to fulfil public interests (Alpa, 1990, Sacco-De Nova, 2004). In other words, the public administration is not exempt from the obligation of good faith; a fortiori, its status gives private individuals even more right, even in the light of the Constitution and the rules of administrative procedure, to fully expect behaviour based on fairness (Nicolussi-Zecchin, 2021).

However, within this framework, the gap with English common law - where the obligation of good faith and precontractual liability are not recognised (Castronovo, 2007) - is narrowed. In actual fact, a different rule is established for procurement and engineering contracts, which are governed by a legislative procedure. In such instances, if specifications have been delivered and client refuses to sign the contract after behaving in a way that led the other party to reasonably believe that it would have been signed, the English courts have sometimes recognised the right to damages.[1] (Castronovo, 2001).

To sum up, at least in Italy, some of the most famous cases of liability for the unjustified breaking off negotiations have concerned public contracts that were not signed even though the adjudication procedures were at an advanced stage (Ilacqua, 2009): for example, cases where only the approval of the contract by the public control authority was missing (Albanese, 2018).

A case in point is a company that took part in a tender for the construction of a major public works project and won the contract, only to have the decision subsequently overturned due to a series of legal errors. According to the Italian Corte di cassazione, this result «undermines the contractor's expectation of fairness in the selection of his tenderer at preliminary and preparatory stage of the contract, where future contractors must not be harmed in their mutual trust in the good faith of the other party». (Cass. 21-11-2011, n. 24,438).

Recently, in Italy, cases of contractual liability of the public administration have also been qualified as cases where the measure adopted is not affected by illegality, it is even perfectly legible, but the behaviour of the public administration has nevertheless breached good faith. In fact, a famous Italian scholar has spoken of an 'equal administrative law', going so far as to postulate the protection of the citizen's trust long before the exercise of public power (Benvenuti, 1996).

Precisely because it damages the confidence of the private party, the Italian Court of Cassation, for example, has held that the legitimate failure to obtain the final approval of the authority for a tender contract already concluded between a private company and the Ministry of Defence is a breach of the obligation of good faith.[2] Similarly, the Italian Consiglio di Stato has held that, even in the case of lawful withdrawal of acts relating to a procedure for works contract (due to the monitoring of the non-availability of financial sources), there may be precontractual liability on the part of public administration, which has acted contrary to the rules of good faith and fairness. In particular, as soon as it became aware of the needed for revoking the decision to conclude the contract, it did not immediately withdraw the previous measures. As a result, the tendering process was

[1] Blackpool & Fidley Aero Club LTD v. Blackpool Borough Council (1990) 1 W.L.R. 1195 C.A.
[2] Cass. 12-7-2016 n. 14,188.

prolonged unnecessarily, leading competing companies to believe that they were well placed to win the contract.[3]

With regard to these cases, which are different from the ones of good faith's breach linked to procedural irregularities on the part of public administration, it has been proposed to use the term pre-action, instead of precontractual, liability. On the one hand, the relationship is in fact intended for the promulgation of a unilateral act by public administration and not for the negotiation of a contract. On the other hand, it is not precisely the legality of provision that disappoints the trust of private parties, but rather the behaviour of the public administrations before the promulgation of the even legal provision (Nicolussi-Zecchin, 2021).

As a result, the rules on precontractual liability cannot be directly applied, whatever the applicable law may be. Rather, it provides a model that can be argued as a general rule: just as each party is obliged to behave in good faith during negotiations of a contract, similarly this assessment is still valid in a relationship aimed at the production of an unilateral act, as in the case of a relationship between public administration and citizen, where the former reserves the power to decide until the last moment whether or not to adopt a public act (Mazzamuto, 2021). Finally, the behaviour to be assessed in terms of good faith is prior to such an act, namely to the decision of the public administration to proceed with a specific act or another, or even not to proceed at all. Therefore, the judgement on these behaviours does not involve any undue interference in the administrative discretion (Nicolussi-Zecchin, 2021).

In conclusion, public administration enjoys a margin of discretion, but it is nevertheless obliged not to mislead the citizen as to the adoption of a provision. Otherwise, there would be a breach of good faith leading to an obligation to compensate the damage suffered by the citizen who relied unnecessarily. This was the only way to avoid the inconsistency of protecting a private person relying on a measure subsequently found to have been unlawful, but not protecting a person legitimately relying on the good faith of a public administration (Comporti, 2020).

In this case, too, the rule of liability will be that of breach of contract, but not for failure to issue the act, which is a legitimate result in terms of administrative discretion, but for breach of reliance on the fact that it would be issued (Nicolussi-Zecchin, 2021). In other words, the features of the relationship with public administration, on whose loyalty the private individual relies on, are emphasised (Venosta, 2021).

As for compensation, again, it would be inconsistent to pay for the consequence of not making a provision. In other words, such a measure cannot be the end of the duty of good faith owed by public administration (Cirillo, 2016). Therefore, only losses and lost profits related to relationship between private party and public administration can be considered, according to the standard of reliance interest, and not damages resulting from the absence of a favourable measure (expectation interest) to which the private party is not properly entitled (Casetta, 2005). For example, as we have said with regard to private megaprojects, the costs unnecessarily incurred and the possible loss of earnings deriving from the lost business.

In addition, again from the point of view of damages, there is a particular point relating to the involvement of a public administration: in fact, the victim of the wrongdoing

[3] Cons. Stato 1-2-2013 n. 633.

is often in a special position with regard to hypothetical lost profits. In the case of public evidence tenders, the participating party cannot normally withdraw its candidature and is therefore effectively prevented from considering other transactions that would deny it of the resources needed to carry out the activity of the tender. In short, behaviour of public administration in breach of the principle of good faith also damages companies' ability to take advantage of other business opportunities. From this point of view, it is impossible to rule out the option of protection, perhaps based on the principle of equality, which might consider this profile of relations between private companies and public administration (Mazzamuto, 2021).

4 Good Faith and Public Body's Behaviour in Private Megaprojects

However, there is another scenario that needs to be further considered in the context of megaprojects, and that is the case where the public administration does not play the role of the client, but rather that of the party whose approval is necessary for the work to start. Consider the case where an administrative authorization is required to carry out a megaproject such as a large complex or a highway.

Well, even in such cases, public administration is obliged to act in good faith, although there is even less question of negotiations or pre-contractual liability. To be more practical, it means to prevent the rise of a legitimate belief in the private party as regards the provision, which will then be disappointed because the public administration, although lawful, takes a different decision (Nicolussi-Zecchin, 2021).

Italian case law has held that a public authority had acted in bad faith in a case where a private company was seeking permission for the construction of a large hotel on its land. There had been "intensive interaction" between such a company and the local authority over a number of years, during which the latter had taken actions which gave rise to a legitimate expectation that a favourable measure would be adopted in favour of the former. However, the permit was ultimately refused. The company, which was not challenging the refusal's legitimacy or the delay, but rather the simple conduct of the administration during the negotiation phase leading up to the decision, claimed damages for the expectation created and then disappointed. And the Italian Corte di cassazione has indeed ruled that in such kind of negotiations, too, public administration does not have a different position from that of private parties: it must therefore also, and even more so in view of Article 97 of the Italian Constitution, comply with the obligation of good faith established by the Article 1337 of the Civil Code.[4] The same principle was later confirmed, albeit in a case unrelated to megaprojects.[5]

In conclusion, the obligation of good faith applies to negotiation of public megaprojects to the extent that it is also binding on those public administrations that are not involved in the megaprojects' contracts but are nevertheless linked by a qualified legal relationship with the future contractors. It follows that, from the point of view of liability, the rules are the same for the scenario of public administration being the client of megaprojects. The damages, on the other hand, will normally be similar to those compensable for breach of good faith in megaprojects between private parties.

[4] Cass. 28-4-2020 n. 8236.

[5] Cass. 15-1-2021 n. 215.

References

1. Albanese A (2003) Violazione di norme imperative e nullità del contratto. Jovene, Napoli
2. Albanese A (2018) Responsabilità precontrattuale, in Le parole del diritto. In: Scritti in onore di Carlo Castronovo, III. Jovene, Napoli, pp 1695–1726
3. Albanese A (2021) Buona fede contratto legge. Eur e dir priv:31–89
4. Allen DK (1990) England, Precontractual liability. In: Hondious EH (ed) Reports to the XIIIth Congress International Academy of Comparative Law. Montreal Canada 18–24 August, 1990. Kluwer, Devemter-Boston, pp 125–143
5. Alpa G (1990) Italy, Precontractual liability. In: Hondious EH (ed) Reports to the XIIIth Congress International Academy of Comparative Law. Montreal Canada 18–24 August 1990. Kluwer, Devemter-Boston, pp 197–204
6. Balestra L (2016) Le regole applicabili alla formazione del contratto internazionale di appalto: standard internazionali e discipline nazionali. Corr giuridico:634–647
7. Benatti F (1963) La responsabilità precontrattuale. Giuffré, Milano
8. Besaiso H, Fenn P, Emsley M, Wright D (2018) A comparison of the suitability of FIDIC and NEC conditions of contract in Palestine. Eng Constr Archit Manag 25(2):241–254
9. Benvenuti F (1996) Disegno dell'Amministrazione Italiana. In: Linee positive e prospettive. Cedam, Padova
10. Cardinale E (2001) Il contratto tipo nel commercio internazionale. In: Griffi P (ed) Manule di diritto commerciale internazionale, U. Milano, Giuffrè, pp 91–107
11. Casetta E (2005) Manuale di diritto amministrativo. Giuffré, Milano
12. Castronovo (1986) L'avventura delle clausole generali. Rev Crit Dir Priv:21–30
13. Castronovo C (2001) Principi di diritto europeo dei contratti. In: Parte I e II, edizione italiana. Giuffrè, Milano
14. Castronovo C (2007) La responsabilità precontrattuale. In: Castronovo C, Mazzamuto S (eds) Manuale di diritto privato europeo, II. Giuffré, Milano
15. Castronovo C (2018) Responsabilità civile. Giuffré, Milano
16. Cirillo GP (2016) La giurisdizione sull'azione risarcitoria autonoma a tutela dell'affidamento sul provvedimento favorevole annullato e l'interesse alla stabilità dell'atto amministrativo. Foro amm:1991–2011
17. Close J (2017) A comparative guide to standard form construction and engineering contracts. Law Brief Publishing, Somerset
18. Comporti GD (2020) La responsabilità relazionale a tutela degli affidamenti dei cittadini. Giur it:2530–2541
19. Confortini M (2017) Clausole negoziali. Utet, Torino
20. Federico A (2018) La riforma dei contratti in Francia - Négociations e obblighi di riservatezza. Giur it:1309–1318
21. Flyvbjerg B (2014) What you should know about megaprojects, and why: an overview. Project Manage Inst 45:6–19
22. Gallo P (2004) La responsabilità precontrattuale: la fattispecie. Riv dir civ:295–325
23. Glover G (2017) Introducing NEC4. Int Q 22:1–5
24. Halson R (2018) Liquidated damages and penalty clauses. Oxford University Press, Oxford
25. Hondious EH (1990) General Report, Precontractual liability. In: Hondious EH (ed) Reports to the XIIIth Congress International Academy of Comparative Law. Montreal Canada 18–24 August 1990. Kluwer, Devemter-Boston, pp 3–28
26. Ilacqua A (2009) La responsabilità precontrattuale della pubblica amministrazione. In: Evoluzioni giurisprudenziali, in Giust amministrativa, pp 418–435
27. Mantovani M (2011) art. 1337. In: Commentario del Codice civile, E. Gabrielli (dir), Dei contratti in generale, E. Navarretta-A. Orestano, artt. 1321–1349. Utet, Milano, pp 440–487

28. Mazzamuto S (2021) La responsabilità contrattuale in senso debole. In: Mazzamuto S (ed) Le tutele contrattuali e il diritto europeo. Scritti per Adolfo di Majo, Napoli, pp 131–166
29. Mengoni L, Scritti I (2011a) In: Castronovo C, Albanese A, Nicolussi A (eds) Metodo e teoria giuridica. Jovene, Napoli
30. Mengoni L, Scritti II (2011b) In: Castronovo C, Albanese A, Nicolussi A (eds) Obbligazioni e negozio. Jovene, Napoli
31. Nicolussi A (2021) Le obbligazioni. Milano, Cedam
32. Nicolussi A, Zecchin F (2021) La natura relazionale della responsabilità «pre-provvedimentale» della pubblica amministrazione. Autorità e affidamento. Eur e dir priv:804–842
33. Piraino F (2015) La buona fede in senso oggettivo. Giappichelli, Torino
34. Roppo V (2011) Il contratto, 2nd edn. Giuffrè, Milano
35. Sacco R, De Nova G (2004) Il contratto, 3rd edn. Utet, Torino
36. Venosta F (2021) «Contatto sociale» e affidamento. Giuffré, Milano
37. Zecchin F (2020) Il danno non patrimoniale dal torto al contratto. Giuffré-Francis Lefebvre, Milano
38. Zimatore A (2010) Note sui cc.dd. accordi di riservatezza, Scritti in onore di Francesco Capriglione. Cedam, Padova, pp 725–730

Mega-Projects and Permanent Establishment: The Relevance of the Negative List Carve-Outs and the Fragmentation of Activities

Paolo Arginelli[1(✉)] and Francesco Reboli[2]

[1] Università Cattolica del Sacro Cuore, Piacenza, Italy
Paolo.Arginelli@unicatt.it
[2] Pirola Pennuto Zei e Associati, Milan, Italy
Francesco.Reboli@studiopirola.com

Abstract. Mega-projects due to their cross-border nature are increasingly influenced by the evolving international tax framework. The successful execution of these projects and, indirectly, the significant positive spillovers effects which can be triggered, hinges on navigating complex legal environments that can either foster or hinder their development. Tax and in general regulatory considerations are crucial as they shape the ecosystem in which such large-scale initiatives can thrive, directly affecting their positive economic outcomes. Given the inherently international nature of mega-projects, the notion of permanent establishment come into play as cornerstone concept in international taxation. This analysis will explore certain nuances of the PE concept within the context of mega-projects, adopting and evolutive view of the relevant provision and stressing the latest amendments that occurred both at international and at Italian level. Understanding the intricacies of PE is therefore essential for stakeholders to navigate international tax issues effectively, ensuring that mega-projects are structured in a way that minimizes legal and financial risks while maximizing their potential for success.

Keywords: International Taxation · Permanent Establishment · Double Tax Conventions · Auxiliary and Preparatory Activities · Negative List · Anti-Fragmentation · BEPS Project

1 Introduction

The question of whether a permanent establishment ("PE") exists is one of the oldest and most frequently encountered issues in tax treaties, frequently addressed by advisers, government officials, and courts. Despite the extensive literature on this subject, the

---

Sections 1, 2, and 3 were written by Paolo Arginelli, while Sect. 4 was written by Francesco Reboli. The authors drafted together the conclusions in Sect. 5. The scientific responsibility of the contribution is shared by the authors.

E. Favari et al. (Eds.): MeRIT 2024, LNCE 600, pp. 108–132, 2025.
https://doi.org/10.1007/978-3-031-92797-3_9

concept of PE and the associated rules become particularly critical in the context of cross-border projects.[1] This holds true for mega-projects, which typically entail complex cross-border activities, involving different entities characterized by diverse value chains and operations. In such cases, international taxation emerges as a strategic consideration to be addressed and duly explored at the initial planning phase, with significant implications for the medium- and long-term feasibility and success of the projects.

Against this background, this Chapter seeks to examine the concept of PE, with a particular focus on the so-called "negative list" in the light of the latest amendments, The negative list delineates a series of activities that are carved-out from the definition of PE, thereby potentially narrowing the PE threshold and consequently limiting the source states taxing powers. In this sense, the chapter aims to delve into the nuances of this provision to highlight the relevant trends and interpretative issues that could be relevant for the implementation of mega-projects.

The analysis is structured into two main parts. The first part examines the notion of PE (Sect. 2) as outlined in Article 5 of the 2017 OECD Model Tax Convention (hereinafter "OECD MC"),[2] providing a glimpse into the historical background of this provision and discussing the original purpose of the PE negative list (Sect. 3.1). The chapter then addresses "preparatory and auxiliary" character as well as the question of whether the activities specifically mentioned in Article 5(3) of the 1963 OECD Draft MC and Article 5(4) of the 1977–2014 OECD MC should be regarded *as a priori* exclusions from PE status (Sect. 3.2). The analysis continues by exploring the changes introduced by the 2017 OECD MC. Furthermore, it examines the application of the "preparatory and auxiliary" character test should in cases involving the fragmentation of connected activities pursuant to the amendments implied by the BEPS Project.[3] The second part focuses on the Italian approach to this provision and how it is addressed both in domestic legislation and in Double Taxation Conventions ("DTCs"). This section also explores the approach adopted by Italian courts, tax authorities, and the trends highlighted in the most recent Italian DTCs to shed light on the evolving Italian stance on this matter. Both parts further discuss how these issues are dealt with under the Multilateral Convention

[1] See, J. Sasseville, A.A. Skaar, *"Is there a PE?"* (IFA General Report) in IFA, IFA Cahiers, 2009, Vol 94A, p. 17 ss.

[2] OECD (2017), Model Tax Convention on Income and on Capital: Condensed Version 2017, OECD Publishing, Paris, https://doi.org/10.1787/mtc_cond-2017-en.

[3] OECD, Action Plan on Base Erosion and Profit Shifting, Paris, OECD Publishing, 2013 https://doi.org/10.1787/9789264202719-en. For a first discussion of the BEPS Project see in literature Y. Brauner, *BEPS: An Interim Evaluation*, World Tax J., 2014, 6, 10 ss.; P. Baker, *Is There a Cure for BEPS?*, Brit. Tax Rev., 2013, 58, 605 ss.; Y. Brauner, *What the BEPS*, Fla. Tax Rev., 2014, 16, 55 ss.; A. Christians, S. E. Shay, *Assessing BEPS: Origins, Standards, and Responses* (General Report), 102A Cahiers de Droit Fiscal International: Assessing BEPS: Origins, Standards, and Responses, Int'l Fiscal Ass'n, 2017; A. P. Dourado, *The Base Erosion and Profit Shifting (BEPS) Initiative under Analysis*, Intertax, 2015, 43, 2 ss.

to Implement Tax Treaty Related Measures to Prevent BEPS ("MLI")[4] and offer an overview of the relevant choices made by the contracting parties to the MLI. Additionally, they provide insights into the practices of national tax courts and authorities. Section 5 concludes.

2 The PE Concept and Its Role Within the International Tax System: General Remarks

In the realm of international tax law, the PE generally plays a pivotal role in determining the rights of a state to tax the profits of an enterprise generated therein by foreign entities. The PE concept establishes a minimum threshold of business presence, below which the source country does not have the jurisdiction to tax the business income of a foreign enterprise. This threshold is generally defined by the requirement of a tangible or physical presence within the jurisdiction.

The origin of such concept is as old as the concept of international taxation itself.[5] Since the early 1920s, the PE threshold has been instrumental in determining whether a particular type of income is subject to taxation in the source country and therefore dividing the taxing powers among jurisdictions involved. According to Article 7(1) of the OECD MC: "*Profits of an enterprise of a Contracting State shall be taxable only in that State unless the enterprise carries on business in the other Contracting State through a permanent establishment situated therein. If the enterprise carries on business as aforesaid, the profits that are attributable to the permanent establishment [...] may be taxed in that other State*" (emphasis added).

[4] See Multilateral Convention to Implement Tax Treaty Related Measures to Prevent Base Erosion and Profit Shifting (24 Nov. 2016), Treaties IBFD. See in literature C. Silberztein, B. Granel & J-B. Tristram, *OECD Multilateral Convention to Prevent BEPS: Implementation Guide and Initial Thoughts*, 24 Intl. Transfer Pricing J. 5 (2017), Journals IBFD; C. HJI Panayi, *International Tax Law Following the OECD/G20 Base Erosion and Profit Shifting Project*, 70 Bull. Intl. Taxn. 11, sec. 5. (2016), Journals IBFD; P.J. Hattingh, *The Multilateral Instrument from a Legal Perspective: What May Be the Challenges?*, 71 Bull. Intl. Taxn. 3/4 (2017), Journals IBFD.

[5] For an historical analysis of the term *see* A.A. Skaar, *Permanent Establishment – Erosion of a Tax Treaty Principle,* Wolters Kluwer, II, 2020, chap. 6. See also J. F. Avery Jones et Al., *The Origins of Concepts and Expressions Used in the OECD Model and their Adoption by States,* Bulletin for International Taxation 2006 (Volume 60), No. 6, p. 220 ff. Accordingly, the expression "permanent establishment" traces back from nineteenth-century German use. In particular such term ("Betriebsstätte") was first used in Prussian tax law in 1885 as one of a number of terms such as agency and place of management to ensure the efficient allocation of tax bases among Prussian municipalities. The term was first used in the tax treaty between Austria/Hungary and Prussia (1899): "*The following are considered as Betriebsstätten: Branch offices, factories, storage facilities, offices, purchase and sale facilities and other business establishments used for the performance of standing enterprises by the entrepreneur himself, business partners, authorised agents or other permanent agents*". The concept of PE was also firstly included in a Model in the first League of Nations Draft (1927). Moreover, in 1958 the OEEC issued its first report, including a permanent establishment definition. This report was eventually incorporated into article 5 of the OECD Draft Model and Commentary (1963).

The PE concepts therefore plays a pivotal role in reconciling two fundamental principles of international taxation: the prerogative of source-based taxation, which allows the country where the income is generated to tax it, and the prerogative of residence-based taxation, where the country of the taxpayer's residence holds taxing rights.[6] As Reimer suggests, the aim underlying the PE principle is threefold: (i) assigning tax revenue to the source state, which ensures that countries hosting economic activities receive a fair share of tax revenues derived from those activities; (ii) maintaining practicability by establishing a minimum threshold, thereby preventing the taxation where the foreign enterprise's presence in the source state is merely incidental or peripheral; (iii) ensuring neutrality and equal treatment, by placing the PE on an equal footing with a resident entities for the purposes of various tax provisions, avoiding discriminatory tax treatment.[7]

The essential characteristics of the PE notion are traceable in the MCs which have been the main driver, historically, to shape the PE notion included both in DTCs as well as in domestic laws.[8] The foundational elements of this concept are rooted in the various MCs developed by international organizations (and single states), with the OECD and UN MCs being the most prominent. These models, along with the interpretive guidance provided by their Commentaries,[9] play a key role in shaping the definition of PE as outlined in Article 5.[10] This provision establishes the essential threshold and criteria that distinguish activities in the source state that are sufficiently rooted in a jurisdiction

[6] To provide certain background, it is worth recalling the longstanding academic debate concerning the existence of boundaries and limitations on the tax jurisdiction of a state. This debate focuses on whether a nexus is required for legislative jurisdiction of a state. See on this topic, J. Kokott, *Chap. 1: Public International Law and Taxation: Nexus and Territoriality* in E. Traversa (Ed.) *Tax Nexus and Jurisdiction in International and EU Law*, IBFD, 2022; S. Gadžo, *The Principle of "Nexus" or "Genuine Link" as a Keystone of International Income Tax Law: A Reappraisal*, 46 Intertax (2018), p. 194 ff. See also P. Baker, Chap. 11: Some Thoughts on Jurisdiction and Nexus in G. Maisto (ed.) Current Tax Treaty Issues: 50th Anniversary of the International Tax Group, IBFD, 2020.

[7] See E. Reimer, *Permanent Establishment*, in E. Reimer and A. Rust (eds.) *Klaus Vogel on Double Taxation Conventions*, Wolters Kluwer, 2022 p. 380, para. 284.

[8] Italy is one such example of a jurisdiction that adopts a domestic concept of PE in line with MCs. Conversely, other jurisdictions do not mirror the MCs in their domestic definitions of PE. A prominent example is the United States, which uses the concept of "trade or business within the United States" under its domestic law, a term that closely resembles the notion of PE. Although this term is not explicitly defined in the U.S. Internal Revenue Code (26 CFR § 1.864–2), several Code provisions and case law provide detailed guidance on its application to specific factual situations. For further discussion on this topic, see M. Herzfeld, *International Taxation in a Nutshell*, West Academic, 2023; W. Choi, United States - Corporate Taxation sec. 7., Country Tax Guides IBFD, 2024.

[9] See OECD (2019), Model Tax Convention on Income and on Capital 2017 (Full Version), OECD Publishing, Paris, https://doi.org/10.1787/g2g972ee-en [hereinafter "2017 OECD Commentary"]; UN (2021) United Nations Model Double Taxation Convention between Developed and Developing Countries, September 2021 https://desapublications.un.org/publications/united-nations-model-double-taxation-convention-between-developed-and-developing-0 [hereinafter "2021 UN MC"].

[10] For the purpose of this Chapter, reference will be mainly made to the OECD MC and its evolution. Specific reference will be also made to the UN MC.

from those that are merely incidental or lack substantial connection to the territory, granting only to the former full taxing power to the profits generated therein.

It goes without saying that the notion of the PE allocates the taxing power of a state, balancing different and (generally) opposed interests of the jurisdictions involved. The notion despite being drafted in the different MCs in a similar way, following typical structure and characteristics, undergoes some deviations in the different existing models that tend either to broader or narrow its scope. Notably, the most prominent distinction is the one between OECD and UN MCs, reflecting the different views of developed and developing states, the former preferring a narrower scope of the PE definition than the latter. This topic will be touched upon incidentally in the next paragraphs.[11]

One final aspect worth highlighting is the ongoing evolution of the concept of PE. The traditional brick-and-mortar economy-based notion of PE has been progressively eroded and challenged by the rapid globalization of the economy, with the rise of emerging economies claiming for a revision of the allocation of taxing powers, as well as by the accelerating pace of digitalization, which have significantly transformed business models.[12] In this context, the evolution of MCs and the parallel BEPS project reflect, at least in part, such phenomenon thereby attempting to adapt the PE concept to contemporary economic reality.[13] However, the unresolved challenges highlighted by the BEPS project illustrate how the concept of PE is not static but requires continuous reassessment and refinement to remain relevant in an increasingly complex and digitized global economy.[14] Therefore, it is crucial to consider the evolving dimensions of the notion of PE in the following analysis to drive the interpretation of such concept.

[11] See in literature A. Binder, V. Wöhrer (eds.), *Special Features of the UN Model Convention*, Linde, 2019; V. Daurer, *Tax Treaties and Developing Countries*, Wolters Kluwer, 2014.

[12] See in literature S. Cipollina, *I redditi nomadi delle società multinazionali nell'economia globalizzata*, Riv. dir. Fin. sc. fin., 2014, I, 21 ss. See also Assonime, Note e Studi no. 17/2016, p. 60 ss.

[13] See on this topic, OECD, Preventing the Artificial Avoidance of Permanent Establishment Status, Action 7–2015 Final Report, OECD Publishing, 2015, https://doi.org/10.1787/9789264241220-en.See in literature, P. Pistone, J. Nogueira & B. Andrade, *The 2019 OECD Proposals for Addressing the Tax Challenges of the Digitalization of the Economy: An Assessment*, ITAXS 2019–2, p. 14; P. Hongler and P. Pistone, *Blueprints for a New PE Nexus to Tax Business Income in the Era of the Digital Economy*, 2015, p. 25. On the nuances of the BEPS Project *see* C. HJI Panayi, *International Tax Law Following the OECD/G20 Base Erosion and Profit Shifting Project,* 70 Bull. Intl. Taxn., 2016, 11.

[14] OECD (2020), *Tax Challenges Arising from Digitalisation – Economic Impact Assessment: Inclusive Framework on BEPS*, OECD/G20 Base Erosion and Profit Shifting Project, OECD Publishing, Paris, https://doi.org/10.1787/0e3cc2d4-en. See in literature: L. Carpentieri, Profili fiscali dell'economia digitale, Giappichelli, 2020.

3 The PE negative list

3.1 Purposes, Effects and General Functioning of the "Negative List"

Article 5(4) of the OECD MC[15] enlist a series of activities characterized by preparatory and auxiliary nature (hereinafter also referred as "negative list"). This provision represents a compelling topic both from a practical perspective as the issues related prevent the source states from taxing these activities, thereby limiting their taxing powers, as well as from a more theoretical standpoint due to the relevant interpretative issues arises by this provision and its evolution over time.

Such a provision has undergone significant changes during the last OECD (and UN) MC revision. In this sense, the BEPS project represented a watershed, spurring a revision of such provision aimed at counteracting base erosion and profit shifting phenomenon.[16] However, the history of the "negative list" is quite longstanding, as the first traces of this provision can be found already in the League of Nations' draft models of the 1940s,[17] and it is possible to identify three key moments in its evolution.[18] The first is found in Article 5(3) of the 1963 Draft OECD MC, which introduced the initial version of the negative list. The second phase spans from 1977 to 2014, during which various models were issued, characterized by minor changes in the wording of Article 5(4) but maintaining overall consistency. The current and third stage is embodied in Article 5(4) of the 2017 OECD MC, which incorporates significant revisions driven by the BEPS project. Article 5(4) of the 2017 OECD MC reads as follow[19]:

[15] In the 1963 OECD Model Convention, the so-called "negative list" was included in Article 5(3). However, in subsequent versions of the MC, this provision was moved to Article 5(4).

[16] OECD, Preventing the Artificial Avoidance of Permanent Establishment Status, Action 7–2015 Final Report, para. 12 (OECD 2015). In this respect, the OECD highlighted that, *"[r]egardless of the original purpose of the exceptions included in subparagraphs a) to d) of paragraph 4, it is important to address situations where these subparagraphs give rise to BEPS concerns*".

[17] Indeed, both the Mexico and the London Models included in article V of their Protocol the following inclusive definition of a PE: The term "permanent establishment" includes head offices, branches, mines and oilwells, plantations, factories, workshops, warehouses, offices, agencies, installations, professional premises and other fixed places of business having a productive character. he Commentary is also helpful to elucidate the reasons for excluding from the notion of PE the above-mentioned places of business, as it affirms that to "apply income tax on an assumed profit in the case of such establishments would easily lead to arbitrary or extra-territorial taxation".[8] Thus, the original purpose of the negative list, which was an expression of the required "productive character" of the places of business, was to foster legal certainty, protect the legitimate expectations of the taxpayers and avoid unfair taxation characterized by very loose territorial connections with the state of (secondary) establishment.

[18] See in literature, P. Arginelli, *Chap. 5: PE Negative List and Fragmentation of Activities Within Groups of Companies* in G. Maisto (ed.) *New Trends in the Definition of Permanent Establishment*, IBFD, 2019, para. 5.2.1.; P. Tarigo, *Diritto Internazionale Tributario - II - Trattati fiscali italiani a confronto,* Giappichelli, 2020, p. 96 ff.

[19] The provision mirrored by the UN MC provides for certain departures, limiting the scope of this carve out. The 2021 UN Model-based negative list deviates from the excluding from paragraph 4 (*i.e,* the "preparatory and auxiliary activities") the "delivery". Therefore, contrary to the OECD MC. a delivery activity might result in a permanent establishment under the UN MC.

"Notwithstanding the preceding provisions of this Article, the term "permanent establishment" shall be deemed not to include:

a) *the use of facilities solely for the purpose of storage, display or delivery of goods or merchandise belonging to the enterprise;*
b) *the maintenance of a stock of goods or merchandise belonging to the enterprise solely for the purpose of storage, display or delivery;*
c) *the maintenance of a stock of goods or merchandise belonging to the enterprise solely for the purpose of processing by another enterprise;*
d) *the maintenance of a fixed place of business solely for the purpose of purchasing goods or merchandise or of collecting information, for the enterprise;*
e) *the maintenance of a fixed place of business solely for the purpose of carrying on, for the enterprise, any other activity;*
f) *the maintenance of a fixed place of business solely for any combination of activities mentioned in subparagraphs a) to e),*

provided that such activity or, in the case of subparagraph f), the overall activity of the fixed place of business, is of a preparatory or auxiliary character".

The purpose of this provision is to exclude a series of place of business from constituting a PE to the extent the facility is "preparatory" or "auxiliary".[20] According to the Commentary on Article 5 this provision enlists a number of cases in which the allocation of profit is difficult due to their remoteness from the actual realization of profits.[21] Indeed, from a practical perspective these difficulties do not outweigh the potential outcome in tax revenue of the source state.[22] Moreover, some of the activities listed in article 5(4) tends to ease cross-border commerce, excluding facilities such as purchasing offices and distributing warehouses.[23]

As highlighted by literature, the decision to include a list of exceptions in the PE definition was clearly not a theoretical or dogmatic choice. Rather, the reason for that inclusion was primarily to ensure legal certainty, avoid unfair (extraterritorial) taxation and foster international trade.[24]

---

[20] See P. Arginelli, *Supra* (n. 18).

[21] See OECD MC Commentaries on Article 5, no. 58. The OECD stated that: "*services [performed] are so remote from the actual realisation of profits that it is difficult to allocate any profit to the fixed place of business in question*".

[22] See E. Reimer, *Supra* (n. 7) p. 376 ff.

[23] It is worth mentioning that the UN Model Convention relies on a narrower Article 5(4) which does not includes facilities and stocks if maintained for the purpose of the delivery of goods or merchandise. Thereby DTCs based upon such model provision do not carve out from the PE scope such facilities, not restricting the taxing power of the source state (i.e., the state where such facilities are located).

[24] See P. Arginelli, *Supra* (n. 18).

The equilibrium between tax certainty and flexibility was found in the coexistence of a closed list of excluded activities and of an open catch-all clause (and, in the 1977 MC, of a combination-of-activities clause) aimed at excluding from the notion of PE also all other activities of a "preparatory or auxiliary character". Nonetheless, it is worth pointing out that the effect of such provision is not merely declaratory as activities that *in concreto* would constitute a PE pursuant the PE general clause set by article 5(1) (i.e., a fixed place of business through which the business of an enterprise is wholly or partly carried on), would be carve out from the PE status. Accordingly, such provision represents a threshold which determine for source state a limitation of its taxing power towards the facility, place of business or and agent. As long as the taxpayer exceeds such a threshold, the taxing power will no longer be limited by such provision and according to article 7 of the MC would be entitled to fully tax the profits attributable to such PE.

3.2 The Preparatory or Auxiliary Character General Requirement

3.2.1 Preparatory or Auxiliary Character

Despite its seeming intuitiveness and its simplification aim, a series of key questions arises when it comes to the interpretation of such provision.

The main interpretative issue concerning Article 5(4) of the OECD MC concerns the "preparatory" or "auxiliary" of the activities encompassed in the negative list. On this point the Commentary to the OECD MC 2017 clarifies that the "*exceptions* [i.e., the carve out from the definition of PE] *only apply if the listed activities have a preparatory or auxiliary character*" and that "*the provisions of paragraph 4 are designed to prevent an enterprise of one State from being taxed in the other State if it only carries on activities of a purely preparatory or auxiliary character in that State*" (emphasis added).[25] However, the nature of this requirement has long been debated in the literature, particularly regarding whether the "preparatory or auxiliary character" of the activity should be regarded as a relative or absolute standard. The former approach seems to be favoured by literature and aligns with the guidance provided by the 2017 OECD MC Commentary on Article 5 (e.g., paragraph 60 ff.).[26] It goes without saying that any activity can be characterized as "preparatory" or "auxiliary" only in respect of another activity and therefore this character cannot be an absolute feature. To put it differently, no activity is intrinsically "preparatory" or "auxiliary". On the contrary, an activity becomes "preparatory" when it is performed in view of another, subsequent, activity. Similarly, an activity becomes "auxiliary" when it is carried out merely in order to support, to serve as an aid to another (main) activity.[27] For instance, delivery services might be an auxiliary function for a

[25] Para. 58 OCED Model: Commentary on Article 5 (2017).

[26] See A.A. Skaar, *Permanent Establishment: Erosion of a Tax Treaty Principle*, Wolters Kluwer, 2020, p. 393 ff.

[27] The OECD Commentary has maintained, since 1977, that the "*decisive criterion is whether or not the activity of the fixed place of business in itself forms an essential and significant part of the activity of the enterprise as a whole*" (see paragraph 24 of the 1977–2014 Commentary and paragraph 59 of the 2017 Commentary). The 2017 Commentary further clarified that "*an activity that has a preparatory character is one that is carried on in contemplation of the carrying on of what constitutes the essential and significant part of the activity of the enterprise*

manufacturing company that integrates delivery as part of its supply chain, but for a logistics or courier company, delivery services constitute the core business activity. The value added by these activities may be comparable, and the potential tax revenue for the source state might be similar in both cases.

This conclusion is strengthened by a contextual interpretation, as the activities explicitly listed in article 5(4)(a) through (e) are limited to cases where they are performed "for the enterprise" or concern "goods or merchandise belonging to the enterprise", which supports the view that no activity is ontologically "auxiliary" or "preparatory", but can be regarded as such when it is performed exclusively in order to support other future or contemporary (main) activities of the same enterprise. Indeed, the very same listed activities, if carried out for the benefit of a third party, or in respect of goods or merchandise belonging to a third party, also only partly (mixed activities), no longer qualify as exceptions to the PE.[28] A possible exception to this rule could be the case of activities performed for the benefit of third parties only in isolated occasions and for an extremely limited economic value, which, as such, could be regarded as ancillary to the activities performed for the enterprise.[29] nevertheless, the Commentary seemingly excludes this possibility considering that a PE would exist whenever the activities performed at the fixed place of business are also carried out for the benefit of third parties.[30]

Such conclusion however triggers an issue of horizontal equity. Indeed, the relative character of the "auxiliary" or "preparatory" nature of the activity makes it theoretically possible that the same activities, where performed by two different enterprises, can lead to different results with regard to the existence of a PE depending on whether those activities qualify as auxiliary or preparatory in respect to the different main activities of the respective enterprises. In this sense, this approach might also result contrary to constitutional principles such as equal treatment and ability to pay.[31] Reimer suggests a combined approach, relying on both standards.[32] The relative standards should be applied

---

as a whole", while "*[a]n activity that has an auxiliary character, on the other hand, generally corresponds to an activity that is carried on to support, without being part of, the essential and significant part of the activity of the enterprise as a whole*". In this regard, the Commentary has also clarified that "*[i]t is unlikely that an activity that requires a significant proportion of the assets or employees of the enterprise could be considered as having an auxiliary character*" (see Para. 60 OECD Model: Commentary on Article 5 (2017)). See P. Arginelli, Supra (n. 18).

[28] On this point the Commentary, which provides that the activities falling into the PE exclusions are only those "*carried on for the enterprise itself. A permanent establishment would therefore exist if such activities were performed on behalf of other enterprises at the same fixed place of business*" (*see* para. 60 OECD Model: Commentary on Article 5 (2017) para. 61).

[29] See in literature P. Arginelli, *Supra* (n. 18).

[30] See, the example at paragraph 61 of the Commentary on Article 5 of the 2017 OECD Commentary.

[31] See in literature P. Arginelli, *Supra (n. 18).*

[32] E. Reimer, *Supra (n. 7).*

initially, but a subsequent application of absolute standards serves as an additional filter.[33] This approach, although reasonable from a policy perspective, does not seem to find support in the wording provided the OECD MC, nor in the Commentary thereon.[34] Moreover, this issue, although possible in theory, should not be too significant in practice, since activities that require significant assets, employees, or investments in order to be carried out would hardly be regarded as not forming an essential and significant part of the activity of the enterprise.[35]

3.2.2 Preparatory or Auxiliary Character of the Activities Listed in Article 5(3) of the 1963 OECD Draft MC and in Article 5(5) of the 1977–2014 OECD MC

As most of the tax treaties in force are based on the 1963 and 1977 OECD MC,[36] a relevant interpretative issue is whether activities falling within the scope of article 5(4)(a)-(d) of the 1977–2014 OECD MC (or article 5(3)(a)-(d) of the 1963 MC) should be deemed a priori PE exclusions, or whether they should be regarded as not giving rise to PEs only where it was proved their preparatory or auxiliary character.

[33] The approach proposed by Reimer relies on two-steps to assess whether the activities of a place of business qualify as "preparatory or auxiliary" under Article 5(4) of the OECD Model. First, the relative standard should be applied by examining whether the activities, compared to the enterprise's overall operations, hold only marginal significance. This analysis should consider the nature, intensity, and relevance of the activity, rather than focusing solely on its contribution to profits or losses. Second, even if activities meet the preparatory or auxiliary criteria, they may still not qualify for the Article 5(4) exemption if the place of business, when viewed independently, demonstrates a sufficient level of size and professional organization. If the place of business would be economically viable as a stand-alone entity, it is presumed not to qualify for the exemption. Conversely, a small business representation with minimal staff, such as six employees without the authority to conclude contracts, would likely not be considered a PE.

[34] Indeed, the 2017 Commentary on Article 5 clarifies that the test to be applied is a relative one, clarifying that "*whether or not the activity of the fixed place of business in itself forms an essential and significant part of the activity of the enterprise as a whole*", and that one relevant factor, for the purpose of establishing whether the activities carried out are of a preparatory or auxiliary character, is "*the overall business activity of the enterprise*" (See Para. 62 *OECD Model*: *Commentary on Article 5* (2017)).

[35] Ibid.

[36] It the case of Italy as it has not yet ratified MLI.

Various arguments have been put forward by scholars, tax authorities, and case law to support these opposing views. It is noteworthy that, considering the changes in wording introduced in the 1977 OECD MC, some of these arguments appear less convincing when applied to the latter group of MCs.

Art. 5(3) 1963 OECD Draft MC	Art. 5(4) 1977 OECD MC[37]
The term "permanent establishment" shall not be deemed to include: *a) the use of facilities solely for the purpose of storage, display or delivery of goods or merchandise belonging to the enterprise;* *b) the maintenance of a stock of goods or merchandise belonging to the enterprise solely for the purpose of storage, display or delivery;* *c) the maintenance of a stock of goods or merchandise belonging to the enterprise solely for the purpose of processing by another enterprise;* *d) the maintenance of a place of business solely for the purchase of goods or merchandise, or for collecting information, for the enterprise;* *e) the maintenance of a place of business solely for the purpose of advertising, for the supply of information, for scientific research or for similar activities which have a preparatory or auxiliary character for the enterprise.*	*Notwithstanding the preceding provisions of this Article, the term "permanent establishment" shall be deemed not to include:* *a) the use of facilities solely for the purpose of storage, display or delivery of goods or merchandise belonging to the enterprise;* *b) the maintenance of a stock of goods or merchandise belonging to the enterprise solely for the purpose of storage, display or delivery;* *c) the maintenance of a stock of goods or merchandise belonging to the enterprise solely for the purpose of processing by another enterprise;* *d) the maintenance of a fixed place of business solely for the purpose of purchasing goods or merchandise or of collecting information, for the enterprise;* *e) the maintenance of a fixed place of business solely for the purpose of carrying on, for the enterprise, any other activity of a preparatory or auxiliary character;* *f) the maintenance of a fixed place of business solely for any combination of activities mentioned in subparagraphs a) to e), provided that the overall activity of the fixed place of business resulting from this combination is of a preparatory or auxiliary character*

The view that the activities listed in subparagraph (a)-(d) should be deemed a priori excluded from the notion of PE finds significant support in the wording of the 1963 OECD Draft MC, in the commentaries as well as in other relevant OECD Materials and

[37] The PE negative list was amended in the 1977 OECD Model, which also renumbered it as paragraph 4 of article 5. Subparagraphs (e) and (f) of the 1977 OECD MC reads as follows:

"e) the maintenance of a fixed place of business solely for the purpose of carrying on, for the enterprise, any other activity of a preparatory or auxiliary character;

f) the maintenance of a fixed place of business solely for any combination of activities mentioned in sub-paragraphs a) to e), provided that the overall activity of the fixed place of business resulting from this combination is of a preparatory or auxiliary character".

The PE negative list provided for in article 5(4) of the 1977 OECD Model remained unchanged through the 2014 OECD Model.

several countries practices. Indeed, the 1963 Draft MC mentions the "preparatory or auxiliary character" of the relevant activities only in subparagraph (e), where it is provided that the term "permanent establishment" shall not be deemed to include "*the maintenance of a fixed place of business solely for the purpose of advertising, for the supply of information, for scientific research or for similar activities which have a preparatory or auxiliary character, for the enterprise*" (emphasis added). The structure suggests that the activities mentioned in subparagraph (e) must be similar to the listed examples and that the "preparatory and auxiliary character" is a specific requirement only for this subparagraph,[38] thereby implying a clear distinction between the activities in subparagraph (e) and those in subparagraphs (a) to (d) being only the activities in subparagraph (e) subject to a case-by-case analysis.

Further support to such view can be traced in the Commentary to 1963 Draft MC, which clarifies that "*[i]n paragraph 3 there is provided a list of specific exceptions which [...] should be excepted from the general rule in order to foster international trade or for convenience of administration*".[39] This conclusion is also supported by the categorical wording of paragraph 12 of the same Commentary, which, with specific regard to a place of business carrying on any of the activities listed in the first part of subparagraph (e) (i.e. advertising, supply of information and scientific research), clarifies that "*the services it performs for its parent enterprise are so far antecedent to the actual realisation of profits by its parent body that no profits can properly be allocated to it; accordingly it does not constitute a taxable unit*".[40]

Additionally, the view that the activities specifically listed in article 5(3) of the 1963 OECD Draft MC should be deemed *a priori* PE exclusions finds indirect support also from the paragraphs of the Commentary dealing with the catch-all clause. On the one hand, paragraph 13 of the Commentary provides that the wording of that clause is "*intended to make it unnecessary to attempt to produce an exhaustive list of exceptions*", a statement that seems to imply that the activities specifically listed in article 5(3) are automatic exceptions to the PE definition. On the other hand, when specifying that it is "*the responsibility of the enterprise to prove that the activities [...] are of a preparatory or auxiliary character*", the Commentary makes exclusive reference to the "other activities" subject to the catch-all clause and not to those expressly listed in the first part of subparagraph (e) and in the previous subparagraphs.[41]

A similar inference holds true also when it comes to the OECD MCs from 1977 to 2014, although with less certainty. The key argument is that the "preparatory and auxiliary" requirement is explicitly mentioned only in subparagraph (e) and in the combination-of-activities clause (subparagraph (f)). Thus, activities in subparagraphs (a) to (d) appear to be automatically excluded from PE status, without requiring proof of their preparatory or auxiliary nature. Nevertheless, the amendments to the wording of the 1977 MC weaken this argument. The removal of specific references to advertising, information supply, and scientific research creates a connection between subparagraph

[38] P. Baker, *Double Taxation Conventions*, Commentary on Art. 5(4) of the OECD Model, para. 5B.20 (Sweet & Maxwell Ltd. 2001).

[39] OECD Income and Capital Draft Model Convention 1963: Commentary on Article 5 para. 5.

[40] Para. 12 OECD Model: Commentary on Article 5 (1963).

[41] Para. 13 OECD Model: Commentary on Article 5 (1963).

(e) and the activities in subparagraphs (a) to (d). Accordingly, the amended proviso may suggest that even these activities might need to meet the "preparatory or auxiliary" requirement to qualify for the PE exception.

The 1977–2014 Commentaries include sections supporting the above conclusion. In particular, the Commentary on Article 5 of the 2014 OECD MC provides that article 5(4) "*lists a number of business activities which are treated as exceptions to the general definition laid down in paragraph 1*",[42] thus apparently excluding that the characterization of the activities listed in subparagraphs (a) thorough (d) as exceptions to the PE definition could be made contingent on the proof of their actual "preparatory or auxiliary character". The Commentary also stress the fact that the "*common feature of these activities is that they are, in general, preparatory or auxiliary activities*", a statement that, in its context, is reasonable to interpret in the sense that the activities expressly listed in subparagraphs (a) through (d), being by their very nature "auxiliary or preparatory", are deemed to be non-rebuttable presumptions of PE exclusions.[43] This latter section of the Commentary, however, does not seem decisive, as the contrary view, although less reasonable, could also be supported based on its face value, i.e. that, since those activities are "auxiliary or preparatory" only in general terms, their actual "auxiliary or preparatory character" should be proved through a case-by-case analysis.[44]

On the contrary, the same Commentary offers an argument in favour of an opposite argument which excludes the existence of a priori preparatory or auxiliary character. In particular, it states that "*the provisions of paragraph 4 are designed to prevent an enterprise of one State from being taxed in the other State, if it carries on in that other State, activities of a purely preparatory or auxiliary character*".[45] This proposition could be construed as implying that the "auxiliary or preparatory" requirement should be met, *in concreto*, also with regard to the activities explicitly listed in subparagraphs (a) through (d).

Finally, it is worth recalling that various OECD materials further support the view that the activities expressly mentioned in the PE negative list of the 1963–2014 OECD MCs should be considered a *priori* excluded from the PE definition. One remarkable example is the 2002 OECD report "*Issues Arising under Article 5 (Permanent Establishment) of the Model Tax Convention*" which clearly states that the activities listed in subparagraphs (a) through (d) are always exempt from the PE status and not subject to an analysis of whether they possess a "preparatory or auxiliary" character.[46] The report emphasizes that these conclusive presumptions were established to provide taxpayers with certainty, ensuring that income from these activities would be taxable, if at all, only in the country of residence. Similarly, the 2004 Final Report titled "*Are the Current Treaty Rules for Taxing Business Profits Appropriate for E-Commerce?*", presented by the Business Profits TAG to the OECD Committee on Fiscal Affairs, underscores that subjecting the activities

[42] See OECD MC 2014, para 21.

[43] Ibid.

[44] For further elements that may support the a priori exclusion argument see P. Arginelli, Supra (n. 18).

[45] See OECD Commentary to art. 5(4) of the 2017 MC, Para 21.

[46] OECD Committee for Fiscal Affairs, Issues Arising under Article 5 (Permanent Establishment) of the Model Tax Convention: 7 November 2002 (OECD 2002).

in subparagraphs (a) through (d) to the "preparatory or auxiliary" requirement would introduce uncertainty,[47] thereby conflicting with the aim of the simplification purposes embedded within the negative list which were intended to foster legal certainty and facilitate international trade. The OECD's Discussion Drafts of 2011 and 2012 were also aligned with this interpretation.[48] Accordingly, the Working Group clarified that the subparagraphs (a) to (d) were not subject to the additional "preparatory or auxiliary" condition, as explicitly required by subparagraphs (e) and (f). Lastly, the 2015 BEPS Final Report on Action 1 indicates that the listed activities should be considered a priori PE exclusions, noting that activities that become significant components of digital businesses and exceed the preparatory or auxiliary threshold would lose their exemption from PE status.[49]

In the light of the above, there are strong arguments supporting the view that the activities explicitly mentioned in the PE negative list provided for by article 5 of the 1963–2014 OECD MCs should be deemed a priori excluded from the notion of PE. These arguments appear stronger with regard to the 1963 OECD Draft MC, due to the specific wording of subparagraph (e) of article 5(3) thereof. However, the wording of Article 5(4)(e) in the 1977–2014 OECD MCs introduces some ambiguity, allowing room for the counterargument that the activities listed in subparagraphs (a) through (d) might only fall outside the PE definition if they are proven, in practice, to possess a "preparatory or auxiliary" character. Despite this potential uncertainty, the historical evolution of the provision, alongside the original objectives of the PE negative list—namely, to ensure legal certainty, prevent unjust (extraterritorial) taxation, and promote international trade—indicates that the same interpretation should reasonably apply to Article 5(4) of the 1977–2014 MCs.

3.3 BEPS Project and the Further Amendments to Such Provision: The Anti-Fragmentation Rule

As a result of Action 7 of the BEPS Project, the 2017 OECD MC incorporated paragraph 4.1, which provides for an anti-fragmentation rule to be applied in connection with article 5(4). The introduction of such provision was aimed at addressing issues of aggregation of different, but cohesive place of business thereby limiting the scope of Article 5(4) OECD and UN MC.[50]

It is worth mentioning that before the BEPS Project had started, the OECD Commentary already considered, although not thoroughly, this issue, focusing on Article 5(4)(f).

[47] Business Profits TAG, *Are the Current Treaty Rules for Taxing Business Profits Appropriate for E-Commerce?* (2004), para. 190.

[48] OECD, *Discussion Draft Interpretation and Application of Article 5 (Permanent Establishment) of the OECD Model Tax Convention* (2011), para 75 and OECD, *Discussion Draft Revised proposals concerning the interpretation and application of article 5 (Permanent Establishment)* (2012), para 76.

[49] OECD, *Addressing the Tax Challenges of the Digital Economy, Action 1*–2015 Final Report, OECD/G20 Base Erosion and Profit Shifting Project: 5 October 2015 para 217.

[50] See E. Reimer, *Supra (n. 7)*; see also V. Dhuldhoya, *The Future of the Permanent Establishment Concept*, 72 Bull. Intl. Taxn. 4a/Special Issue (2018), Journal Articles & Opinion Pieces IBFD (accessed 3 Oct. 2024).

The Commentary on Article 5 of the 1977 OECD MC, for instance, when dealing with the case of combined activities under (then) new subparagraph (f), provided that the latter subparagraph "*is of no importance in a case where an enterprise maintains several fixed places of management within the meaning of sub-paragraph a) to e) provided that they are separated from each other locally and organisationally, as in such a case each place of business has to be viewed separately and in isolation for deciding the question whether or not a permanent establishment exists*". The OECD hence regarded the requisites of "geographical unity" and "organisational connection" as alternative requirements for the purpose of applying the negative list test, so that the meeting of either requisite sufficed for triggering the application of that test to the combined activities. However, this approach appears purported at the mere interpretation of article 5(4)(f) of the OECD MC and not concerned with its application in cases of abuse. Nevertheless, the 2003 amendments to the OECD Commentary considered an anti-abuse perspective.[51] Accordingly with reference to article 5(4)(f), the 2003 Commentary added a definition of "organisationally connected" places of business, according to which "*they each perform in a Contracting State complementary functions such as receiving and storing goods in one place, distributing those goods through another etc*".[52] In addition, the 2003 Commentary highlighted the anti-abusive effect of the "organisational connection" requirement, by stating that an "*enterprise cannot fragment a cohesive operating business into several small operations in order to argue that each is merely engaged in a preparatory or auxiliary activity*".[53]

Against this background, the "anti-fragmentation" rule is not yet widespread as most bilateral DTCs lack a clause corresponding to Article 5(4.1) OECD and UN MC. Some countries such Finland, Luxembourg, Sweden and Switzerland as have also reserved the right not to include new paragraph 4.1 in their tax treaties.[54] Nevertheless, this might differ from a perspective standpoint with a progressively broader implementation of Article 13(4) of the MLI by all Signatory Parties as well as for newly signed DTCs which, at least for Italy, tends to include such provision (*see* further para 4.3). Article 5(4.1) reads as follows:

"*Paragraph 4 shall not apply to a fixed place of business that is used or maintained by an enterprise if the same enterprise or a closely related enterprise carries on business activities at the same place or at another place in the same Contracting State and*

a) *that place or other place constitutes a permanent establishment for the enterprise or the closely related enterprise under the provisions of this Article, or*

[51] See L. De Broe, *International Tax Planning and Prevention of Abuse: A Study under Domestic Tax Law, Tax Treaties and EC Law in Relation to Conduit and Base Companies*, IBFD Doctoral Series, vol. 14, p. 381 et seq. (IBFD 2008); B.J. Arnold, *Tax Treaties and Tax Avoidance: The 2003 Revisions to the Commentary to the OECD Model*, 58 Bull. Intl. Fiscal Docn. 6, p. 246 (2004), Journal Articles & Papers IBFD and A. Martín Jiménez, *The 2003 Revision of the OECD Commentaries on the Improper Use of Tax Treaties: A Case for the Declining Effect of the OECD Commentaries?*, 58 Bull. Intl. Fiscal Docn. 1, p. 18 (2004).

[52] Para. 27.1 *OECD Model: Commentary on Article 5* (2003).

[53] Ibid.

[54] Para. 212 OECD Model: Commentary on Article 5 (2017).

b) *the overall activity resulting from the combination of the activities carried on by the two enterprises at the same place, or by the same enterprise or closely related enterprises at the two places, is not of a preparatory or auxiliary character, provided that the business activities carried on by the two enterprises at the same place, or by the same enterprise or closely related enterprises at the two places, constitute complementary functions that are part of a cohesive business operation*".

Although the language followed by Article 5(4.1) is new, it could be seen as declaratory. In fact, it encompasses a rule that brings into the text of the Model the "organisational connection" requirement already demanded by the 1977 Commentary regarding article 5(4)(f). However, this provision shows a rather innovative character going beyond the pure anti-abusive approach. Despite a declared anti-abusive objective,[55] the paragraph is structured as an objective rule, that applies regardless the actual purpose underlying the fragmentation of the activities between related enterprises.[56]

Therefore, Paragraph 4.1 has the effect to preclude the application of paragraph 4, thereby preventing the possibility to benefit from the PE exceptions. However, the anti-fragmentation rule requires two conditions be triggered.

First, the activities carried out by the two (or more) enterprises at the same place of business, or by the same enterprise or closely related enterprises at the two (or more) places of business, are subject to article 5(4.1) only where they constitute "*complementary functions that are part of a cohesive business operation*".

Second, paragraph 4.1 applies only where (i) the "complementary functions" performed in the source state are not, where considered in combination, of a preparatory or auxiliary character, or (ii) one or more of the places of business where the "complementary functions" are performed already constitute a PE under the provisions of article 5. In other words, the latter provision (ii) presumes, without any possibility to provide to the contrary, that the combination of activities has not a "preparatory or auxiliary character" for the sole fact that one (or more) of the places of business through which those activities are performed constitutes a PE under article 5.

More broadly, the term "permanent establishment" are interpreted by the Commentary as not being "*limited to situations where a resident of one Contracting State uses*

[55] The Commentary clarifies that this provision is aimed at preventing "*a group of closely related enterprises from fragmenting a cohesive business operation into several small operations in order to argue that each is merely engaged in a preparatory or auxiliary activity*".

[56] The notion of closely related enterprise is defined by Article 5(8) of the 2017 OECD Model, which reads as follow: "*For the purposes of this Article, a person or enterprise is closely related to an enterprise if, based on all the relevant facts and circumstances, one has control of the other or both are under the control of the same persons or enterprises. In any case, a person or enterprise shall be considered to be closely related to an enterprise if one possesses directly or indirectly more than 50 per cent of the beneficial interest in the other (or, in the case of a company, more than 50 per cent of the aggregate vote and value of the company's shares or of the beneficial equity interest in the company) or if another person or enterprise possesses directly or indirectly more than 50 per cent of the beneficial interest (or, in the case of a company, more than 50 per cent of the aggregate vote and value of the company's shares or of the beneficial equity interest in the company) in the person and the enterprise or in the two enterprises*".

or maintains a fixed place of business in the other State; it applies equally where an enterprise of one State uses or maintains a fixed place of business in that same State". Therefore article 5(4.1) encompasses cases where an enterprise resident of the source states carries out in that state activities that are complementary to those performed in the same state by a non-resident related enterprise, since the former enterprise will usually have a PE in its state of residence. Similarly, article 5(4.1) applies to an enterprise resident of the other contracting state (R) even if its closely related enterprises are resident of third (non-contracting) states, provided that the latter enterprises carry out part of their activities through a fixed place of business or PE in the source (contracting) state. In this respect, the existence of the fixed place of business or PE of the closely related enterprises in the source state must be ascertained based on the definition encompassed in article 5 of the tax treaty in force between the source state and state R.[57]

3.4 Interim Conclusion

Based on the analysis presented in the preceding sections, several conclusions can be drawn regarding the development and the interpretation of Article 5(4).

The inclusion of a list of exceptions to the definition of a PE was primarily driven by practical concerns, such as ensuring legal certainty, avoiding extraterritorial taxation, and fostering international trade.

Moreover, the wording of the 1963–2014 OECD MC and their Commentaries supports the interpretation that the activities explicitly mentioned in the PE negative list were intended to be *a priori* excluded from the definition of PE. This interpretation is particularly strong for the 1963 OECD Draft MC, though later versions (1977–2014) introduced room for debate, particularly with respect to whether the listed activities should be excluded only when proven to be "preparatory or auxiliary" in practice. On the contrary, the 2017 amendments introduced by the OECD as part of the BEPS Action 7 were not intended to clarify earlier versions of Article 5(4) but rather to create a new approach.

It has been also stressed that the "preparatory or auxiliary" nature should be understood as a relative concept, rather than an absolute one. The evaluation of whether activities in the source state qualify as such must take into account the broader context of the non-resident enterprise's overall business operations. This assessment should consider complementary functions carried out by related enterprises, particularly when those functions are part of a cohesive business operation.

In addition, the decisive element to determine "preparatory or auxiliary" character should be whether the fixed place of business plays an essential and significant role in the enterprise's overall activities. Only activities conducted exclusively for the enterprise, and not for third parties, can qualify for such carve-out. Mixed activities that benefit both the enterprise and third parties should not be considered "preparatory or auxiliary", except in cases where the benefit to third parties is incidental and economically insignificant. Furthermore, activities that directly serve the enterprise's primary purpose, including managerial and contract negotiation activities, cannot be considered "preparatory or auxiliary".

[57] P. Arginelli, *Supra* (n. 18).

Lastly, the issue of activity fragmentation, particularly within multinational groups, has been a concern since the 1977 OECD MC amendments. The OECD's approach, confirmed by the 2003 Commentary and the BEPS Action 7 reforms, emphasizes that artificially fragmented activities across related enterprises should be recombined for the purpose of applying the "preparatory or auxiliary" test. The introduction of Article 5(4.1) in the 2017 OECD MC and its inclusion in the MLI reflect a broader effort to prevent tax avoidance through fragmentation. This new anti-fragmentation rule applies objectively, regardless of the intention behind the division of activities. However, resistance to these changes persists, with some countries opting out of the MLI provisions that address fragmentation.

Nonetheless, this provision could have a notable impact, particularly in terms of tax certainty and administrative burden, on mega-projects involving highly integrated corporate groups. As the inclusion of such provisions becomes increasingly common in new treaties or through subsequent amendments, it may represent a critical factor to consider in the planning and implementation of these projects. Ensuring compliance with evolving tax regulations will be essential to mitigating risks and promoting the success of large-scale, cross-border projects.

4 PE and Negative Under Italian Legislation: Recent Trends

4.1 The Relationship Between the Domestic Provision and the Tax Treaty Rules

The PE definition was originally introduced into Italian domestic tax law at article 162 of the Income Tax Code ("ITC") by Legislative Decree no. 344 of 12 December 2003, with effect from 1 January 2004.[58] This PE definition is valid for corporate for personal income tax purposes (*Imposta sul reddito delle persone fisiche*, "IRPEF"), as well as for corporate income tax (*Imposta sul reddito delle società*, "IRES") and regional tax on productive activities (*Imposta regionale sulle attività produttive*, "IRAP") purposes, while for value added tax ("VAT") purposes a separate definition applies.[59]

[58] For the previous tax periods different provision applied (i.e., article 20 of the old ITC). Nonetheless no definition were provided under domestic law, which referred to the OECD materials. The PE concept, however, was already present in the domestic and treaty tax landscape way before 2004. In particular, the Italian revenue agency and courts worked developed a PE concept even before the inclusion of the PE definition into domestic tax law, by making reference to the notion of PE outlined in the OECD Model and generally included in Italian tax treaties. See Agenzia delle Entrate, Circular Letter of 16 June 2004, no. 25/E: "*Tenuto conto che la Legge delega (art. 4, comma 1, lett. a)) imponeva al legislatore l'utilizzazione di criteri desumibili dagli accordi internazionali contro le doppie imposizioni, il dettato dell'art. 162 riprende sostanzialmente il testo più recente dell'art. 5 del Modello di Convenzione OCSE, discostandosene solo in alcuni punti, in considerazione delle interpretazioni contenute nel commentario a detto modello concordate tra gli Stati membri*".

[59] For VAT purposes, the different concept of "fixed establishment" becomes relevant. This concept is defined by article 11 of European Council Regulation 282/2011, including any establishment, other than the place of establishment of a business, sufficiently permanent and suitably structured in terms of human and technical resources to enable it to receive and use the services supplied to it for its own needs.

The PE definition mainly mirrored article 5 of the 1977 OECD MC, albeit with certain departures. However, Article 1(1010) of Law no. 205 of 27 December 2017 ("2018 Budget Law") redrafted article 162 of the ITC in order to align the domestic PE definition with the recommendations made by the OECD in the BEPS Project which were reflected in both the Multilateral Instrument and in the latest version of OECD MC (2017).

Among the plethora of significant amendments implemented by the 2018 Budget Law,[60] for the purpose of this Chapter it is worth mentioning that negative list modified by redrafting article 162(4) of the ITC and adding new article 162(4-*bis*) thereto, in order to ensure that not only the combination of the activities included in the negative list but also each single activity taken individually must have a preparatory or auxiliary character in order not to give rise to a PE. In addition, the 2018 Budget Law also introduced the anti-fragmentation rule at article 162(5) of the ITC, aimed at countering the fragmentation of the activities carried out on the Italian territory by closely related enterprises with a view to avoid the creation of a PE by exploiting the exclusions of the negative list.

4.2 The Preparatory and Auxiliary Activity Exclusion Within the Domestic Context

As anticipated in the previous subparagraph, Italy's domestic provision on preparatory and auxiliary activities was lastly revised by the 2018 Budget Law, which amended Article 162(4) and introduced Article 162(4-bis) to the ITC to align with the international developments of the 2017 OECD MC.[61] Before these changes, the domestic "negative list" largely mirrored Article 5(4) of the 2014 OECD MC. Under the current version, each listed activity must now be assessed for its preparatory or auxiliary character to determine whether it triggers a PE. This shift, which apparent in the light of the new wording of Articl 162(4) of the ITC, means that the listed activities are no longer automatically

[60] The 2018 Budget Law provided various amendments to the domestic rules, among of which it: (i) introduced a new item in the list of fixed places of business which generally constitute PEs (the "positive list"); (ii) amended the preparatory and auxiliary activities list (the "negative list"); (iii) introduced an "anti-fragmentation rule"; (iv) amended the new definition of agency PE; and (v) introduced a definition of "person closely related to an enterprise"; (vi) adding new paragraph f-bis) to article 162(2) of the ITC.

[61] Article 162(4) and (4-bis) of the ITC reads as follows: 4. *Notwithstanding paragraphs 1 to 3, the term "permanent establishment" shall be deemed not to include: a) the use of facilities solely for the purpose of storage, display or delivery of goods or merchandise belonging to the enterprise; b) the availability of goods or merchandise belonging to the enterprise stored solely for the purpose of storage, display or delivery; c) the availability of goods or merchandise belonging to the enterprise stored solely for the purpose of processing by another enterprise; d) the availability of a fixed place of business used solely for the purpose of purchasing goods or merchandise or of collecting information for the enterprise; e) the availability of a fixed place of business used solely for the purpose of carrying on, for the enterprise, any other activity; f) the availability of a fixed place of business used solely for any combination of activities mentioned in letters a) to e). 4-bis. The provisions of paragraph 4 shall apply on condition that the activities mentioned in a) to e) or, in the cases of f), the overall activity of the fixed place of business is of a preparatory or auxiliary character.*

considered preparatory or auxiliary, requiring a case-by-case analysis based on factual circumstances.[62]

Notably, under the previous PE definition, which adhered to the 2014 OECD MC, it was unclear whether Italy interpreted the activities in the negative list as inherently excluded from PE status or only excluded if they had a preparatory or auxiliary nature in practice. Specifically, the wording of Article 162(4) (i.e., the version applicable before the 2018 Budget Law amendments) stated that "*a fixed place of business shall, in any case, not be deemed to constitute a permanent establishment [...]*" therefore supporting the former interpretation. Moreover, in Resolution no. 4/2017,[63] the ITA seemed to exclude the need for case-by-case analysis, emphasizing that the activities listed in Article 5(4) of the OECD MC were by nature preparatory or auxiliary, and thus did not constitute an essential part of the non-resident enterprise's business.[64] A contrary position, as highlighted by literature, has been put forward in recent years by some local offices of the tax authorities in the course of the tax audits carried out in respect of non-resident digital economy companies.[65]

As for the meaning of "preparatory" and "auxiliary", the Italian Tax Police clarified that in compliance with the OECD approach, an activity shall be considered preparatory if it is instrumental to an essential and significant part of the activity carried out by the enterprise, i.e. of its core business, and it is part of it, whereas it shall be considered auxiliary if it supports the core business of the enterprise without being part of it.[66] Moreover, the ITA considered that the permanent establishment, in order to qualify as such, must conduct a full cycle of business activity with its own economic result, independent of that achieved by the head office.[67]

These concepts were further shaped and interpreted by relevant case law. Among others, the Italian Supreme Court in 2015[68] addressed Article 5 of the Italy-Luxembourg Income and Capital Tax Treaty (1981) in a case involving a toll manufacturing Italian company serving a Luxembourg-based associated company. The Italian company stored and delivered finished goods, while the Luxembourg company managed marketing and sales. The Italian tax authorities claimed that the Luxembourg company had a hidden PE in Italy, citing the Italian company's facilities, employee involvement in order collection, and the use of an Italian-based server. The Court, upholding lower court decisions, ruled

62 See also Italia Tax Police, Circular Letter no. 114153 of 13 Apr. 2018, Attachment no. 2, p. 32.

63 The case concerned the establishment of a logistic hub in Italy. The logistic hub carried out various activities such as storage, exposition, display and delivery of finished goods.

64 In particular it was stated that "a*s clarified by paragraph 4 of the Commentary on Article 5 of the OECD Model, these are activities which are exclusively of a preparatory and auxiliary character, which means activities that do not per se constitute an essential and significant part of the activity of the non-resident enterprise*" (Author's translation). See also resolution of the Italian Ministry of Finance, Resolution No. 12/533 of July 21, 1983 dealing with article 5(3)(d) of the DTC between Italy and Switzerland and which address the procurement activity.

65 See Assonime, Circular Letter no. 15/2018, p. 55 ff.

66 See, Italian Tax Police, Circular Letter no. 114153 of 13 Apr. 2018, Attachment no. 2, p. 32.

67 See Italian Ministry of Finance, Resolution no. 9/2398 of 10 february 1983. See also Italian Ministry of Finance, Resolution no. 12/533 of July 21, 1983. See in literature M. Leo, Le imposte sul reddito nel testo unico, Milano, 2024, 3053.

68 Decision no. 5649 of 20 March 2015.

that the technological platform and order-collecting employees did not constitute a PE. The judgment suggests that the storage and delivery of goods were not viewed as core activities of the Luxembourg company.[69]

Furthermore, the Italian Supreme Court has emphasized that when the activities of a place of business involve financing functions or the provision of services to other companies, even if they belong to the same group as the parent company, these activities cannot be classified as merely preparatory or auxiliary. Instead, such activities are considered to pursue an independent economic outcome distinct from that of the head office. [70]

In addition, article 162(5) includes the so-called "anti-fragmentation rule", which mirrors the provision set forth by article 5(4.1) OECD MC 2017 (as well as article 13(4) of the MLI). Article 162(5) of the ITC is aimed at countering the fragmentation of the activities carried on in the territory of the state by closely related enterprises, with a view to abuse of the negative list exclusions. In this sense, the Italian Tax Police, qualified such provision as having an anti-abuse nature.[71]

Pursuant to the domestic anti-fragmentation rule, the "negative list" exemptions shall not apply to a fixed place of business provided that: (i) the non-resident enterprise or a closely related enterprise carries out business activities at the same location or at a distinct location in Italy; (ii) such location(s) constitute(s) a PE for either enterprise under article 162 of the ITC, or the overall activity resulting from the combination of the activities carried out at the same location, or at the two distinct locations, are not preparatory or auxiliary; and (iii) the business activities carried out at the same location, or at the two distinct locations, are complementary and part of a cohesive business operation.[72] In this sense, the Italian Tax Police, stated that, in case of tax audits, it is necessary to consider each activity carried out by the group in Italy, in order to combine, if integrated and part of a cohesive business, the economic functions and processes carried out by different companies or at different locations, thereby assessing the auxiliary and the preparatory character to the combined activities.[73]

It is worth mentioning that Assonime[74] suggested that the introduction of the anti-fragmentation rule in the PE domestic definition should be viewed as legislative recognition of principles previously established by the Supreme Court in cases such as *Philip Morris,*[75] *SMI San Marino case*[76] *and CEPU case.*[77] Specifically, in the latter case, the Court recognized the existence of a PE for a San Marino-based company, which had divided its business operations in Italy across several legally independent, yet economically integrated entities. The Court ruled that, when determining whether non-resident

[69] See T. Galluccio, *Chap. 17: Italy* in G. Maisto (ed.), New Trends in the Definition of Permanent Establishment, IBFD, 2019, p. 16.

[70] See Italian Supreme Court, Ord. no. 21693 of 8 October 2020, concerning tax periods between 1999 and 2006.

[71] See, Italian Tax Police, Circular Letter no. 114153 of 13 Apr. 2018, Attachment no. 2, p. 29.

[72] See T. Galluccio, *Supra (n. 69)*, p. 15.

[73] Ibid.

[74] See Assonime, Note e Studi no. 17/2016, p. 62 ff.

[75] Italian Supreme Court, Decision, no. 7682 of 25 May 2002.

[76] Italian Supreme Court, Decision no. 1118 of 17 January 2013.

[77] Italian Supreme Court, Decision no. 20597 of 7 October 2011.

companies have PEs in Italy, it is irrelevant whether the activities are conducted by multiple separate entities or a single one. On the contrary, facts and circumstances must be considered in order to understand whether these entities operate as parts of a unified, economically integrated structure aimed at fulfilling the business objectives of non-resident companies in Italy. Accordingly, such principle would be already applicable in Italy, independent of any future changes to the Italian Conventions on Taxation, given the precedent established by the Supreme Court.[78]

4.3 Tax Treaty Practice

Within the domestic legislative framework, DTCs once implemented by a ratification law, acquire the value of a primary source, pursuant to Articles 10(1) and 117 of the Constitution, as reiterated, in tax matters, also by Art. 75 of Presidential Decree No. 600 of 1973 ("*international agreements made enforceable in Italy remain unaffected by domestic provisions concerning income taxes*") and by Art. 169 of the ITC (according to which, the provisions contained therein "*apply, if more favourable to the taxpayer, even in derogation of international agreements against double taxation*").[79] According to such provisions, as well as to case law[80] the DTCs, due to the specialty character prevail over the corresponding domestic provision.

Against this background, Italian tax treaty practice concerning preparatory and auxiliary activities predominantly aligns with Article 5(3) of the 1963 OECD Draft MC.[81] As a result, the majority of Italy's tax treaties do not incorporate three key provisions: first, the maintenance of a fixed place of business for any combination of excluded activities, which was introduced later in paragraph (f) of the 1977 OECD MC; second, the anti-fragmentation rule, which was included only in the 2017 OECD MC and third the clarification also introduce with the 2017 OECD MC concerning the auxiliary ad preparatory character.[82]

---

[78] See Assonime, Note e Studi no. 17/2016, p. 62 ff.

[79] See Italian Supreme Court. Decision no. 30140 of 20 November 2020.

[80] See, among others, Italian Supreme Court. Decision no. 1138 of 19 January 2009, Decision no. 14474 of 15 July 2016, Decision no. 23984 of 24 November 2016. See, more recently, Italian Supreme Court. Decision no. 36679 of 14 December 2022.

[81] Among those DTCs, it is worth mentioning the ones stipulated with the following jurisdictions: Albania, Armenia, Australia, Austria, Belgium, Belarus, Bosnia Herzegovina, Brazil, Bulgaria, Congo, Croatia, Cyprus, Czech Republic, Ecuador, Egypt, Ethiopia, Finland, France, Georgia, Germany, Ghana, Greece, Hungary, Iceland, Ireland, Israel, Japan, Jordan, Kazakhstan, Kuwait, Lebanon, Luxembourg, Macedonia, Malaysia, Malta, Mauritius, Moldova, Morocco, Mozambique, Netherlands, New Zealand, Norway, Oman, Philippines, Poland, Portugal, Qatar, Russian Federation, San Marino, Saudi Arabia, Senegal, Serbia and Montenegro, Singapore, Slovakia, Slovenia, South Africa, Spain, Sri Lanka, Sweden, Switzerland, Syria, Tanzania, Thailand, Trinidad and Tobago, Tunisia, Turkey, Ukraine, Uganda, United Arab Emirates, United Kingdom, Uzbekistan, Venezuela, Zambia.

[82] See in literature, P. Tarigo, *Diritto internazionale tributario*, Vol. II, Trattati fiscali italiani a confronto, Torino, 2020, p. 96 ff. See also, T. Galluccio, *Supra* (n. 69); P. Tarigo, *L'influenza sulle più recenti convenzioni italiane dei diversi modelli-tipo di trattato contro le doppie imposizioni*, Rass. Trib. 2022.

A smaller subset of Italian tax treaties deviates from this standard, particularly regarding paragraph (e), which follows the wording of Article 5(4) of the 1977 OECD MC (i.e., "the maintenance of a fixed place of business solely for the purpose of carrying on, for the enterprise, any other activity of a preparatory or auxiliary character"). Notably, these hybrid treaties do not include Article 5(4)(f) of the 1977 OECD MC.[83] However, a limited number of treaties more closely reflect the provisions of the 1977 OECD MC.[84]

Some further relevant departures from this practice can be found in Italy's tax treaties with developing countries, which often rely on the UN MC.[85] These variations reflect broader differences between the two models, particularly in their approach to the allocation of taxing rights to source countries.

When it comes to the most recent DTCs, although there is a general alignment with Article 5(4) of the 2017 OECD MC, notable deviations exist. For instance, the Chile-Italy DTC adheres to the BEPS standards by requiring that all listed activities meet the additional condition of being "preparatory or auxiliary" to qualify for the exemption, and it incorporates the anti-fragmentation rule. However, it does not fully reflect article 5(4) of the OECD MC (2017), as it omits any reference to "*any other activity of a preparatory or auxiliary character*" in paragraph (e) and the combination of activities clause in paragraph (f). It is worth mentioning that some of the most recent DTCs follow closely article 5(4) of the OECD MC 2017, without the above deviations.[86]Conversely, other recent Italian DTCs with countries such as Barbados, Romania, Jamaica, and Taiwan still follow the 2014 version of the OECD MC, albeit with some deviations.[87]

Regarding Italy's position on MLI, the country opted for Option A under Article 13(1) and chose not to opt out of Article 13(4), ensuring that Italian tax treaties will include the anti-fragmentation rule. However, this option (which is not a minimum standard) requires requires that the other state has also opted for an identical choice. If, on the other hand, if the other country chooses a different option (B) or exercises a reservation, Option A will not apply.

Considering the above, therefore most of the Italian DTCs aligning with the 1963 OECD MC should suggest that the auxiliary or preparatory character should be considered as *a priori* element for the activities listed in the "negative list". Furthermore,

[83] These DTCs include those stipulated with: Azerbaijan, China (old version, signed in 1986), Korea (Rep.), Denmark, Estonia, Latvia and Mexico. Among this second group, it should also be included the DTCs stipulated with Argentina, Canada and Ivory Coast.

[84] DTCs stipulated with Barbados, Hong-Kong, Panama, Romania.

[85] DTCs stipulated with Algeria, Bangladesh, India, Indonesia, Pakistan and Vietnam.

[86] Colombia, Liechtenstein and Uruguay. In addition, on 11 September 2024, the Italian Senate approved a draft law ratifying the DTC with Kosovo, signed in 2021, which aligns with the latter treaties in terms of Article 5(4).

[87] It is also the case for the DTC with China, signed in 2019, which is expected to replace the current DTC, in force since 1986. The language in Article 5(4) of the new DTC mirrors that of the 2014 OECD MC, excluding the amended subparagraph (f) and the anti-fragmentation rule. This provision aligns with the MLI approach followed by China. On 11 September 2024, the Italian Senate approved a draft law ratifying this DTC. This approach is consistent with the DTC signed with Libya in 2009 (with an exchange of notes signed in 2014), for which, on the same date, the Senate also approved a draft law ratifying the agreement.

despite the commentaries on OECD 2014 MC allowed some room for alternative interpretations (see supra Sect. 3), it is worth mentioning that any changes to the provisions of the OECD MC (such as paras 4, 4.1, 5, 6 and 8 of Art. 5 following BEPS Action 7) apply only to future treaties concluded based on the updated standard, and not to those already in force. The 2017 Commentary to Art. 5 clarifies the forward-looking status of the amendments to the Commentary concerning para. 4.1 of Art. 5 (Anti-fragmentation rule) and the amendments to paragraphs 4 (Ancillary and preparatory activities). Consequently, these changes to the Commentary do not affect the interpretation of previous provisions in the older MCs where such amendments were not yet included. This principle aligns also with the recent case law of the Italian supreme court which seemingly favours an "ambulatory approach". Nonetheless a limit to such an approach is set considering that changes to the Commentary deriving from amendments to the OECD MC are not relevant to the interpretation of previously concluded DTCs if the provisions substantially differ from the amended provisions of the OECD MC.[88]

4.4 Interim Conclusion

In the light of the above, Italy has amended the domestic provision to align with international standards. However, since many treaties are based upon pre-BEPS MCs, the domestic provision is not effective, since the DTCs being more favourable take precedence in accordance with art. 169 of the ITC. However, these elements constitute a multi-layered scenario that taxpayers, and especially foreign taxpayers have to face. In this sense, the complexities are enhanced by the plethora of rules and their interpretation that overlap.

The amendment introduced in the 2017 OECD MC concerning the second part of Article 5(4)(f) applies only to treaties incorporating this specific wording. For example, the Italy-Uruguay Convention includes this clarification, making Article 162(4-bis) fully applicable. In contrast, treaties following the pre-2017 OECD MC, which do not include this condition, may still be interpreted under the older framework where subparagraphs (a) to (d) were considered automatically "exempt".

In conclusion, the 2017 OECD MC introduced a significant clarification by requiring that "exempt" activities be of a "preparatory or auxiliary" nature, but this applies only to treaties explicitly incorporating this language. Conventions, such as the Italy-Uruguay DTC, which include this new wording, allow the full application of Article 162, paragraph 4-bis. In contrast, treaties that follow the pre-2017 OECD MC, which lack this condition, may still be interpreted under the older framework where certain activities (subparagraphs (a) to (d)) were automatically deemed "exempt".

Although the 2014 OECD Commentary already hinted at some of the concepts later formalized in 2017, the new wording represents an innovation that does not apply retroactively to treaties lacking this provision. For treaties covered by the MLI, the new wording will be effective at the moment when (and in case of) both treaty partners

[88] See Italian Supreme Court, decision no. 36679 of 14 December 2022 with regard to the interpretation of Article 5(4) of the Italy-Switzerland DTC. See in literature G. Maisto, C. Silvani, *Italy: Application of Later Versions of the OECD Commentary to Older Double Tax Treaties* in G.W. Kofler et al. (eds.), *Tax Treaty Case Law around the Globe*, 2023, IBFD, 2024.

adopt the same option. However, for uncovered treaties that retain the old wording, the exemption status will continue to operate under the prior interpretation, limiting the applicability of Italy's domestic rules.

5 Concluding Remarks

The analysis presented in the preceding sections reveals a highly complex scenario. The challenges in identifying the constituent elements of a PE are compounded by the intricate task of determining which set of rules applies. This complexity arises from the significant overlaps between domestic legislation, relevant tax treaties, the OECD MC on which those treaties are based, and the corresponding Commentary. Moreover, the introduction of the MLI, if signed by both contracting states, adds further complexity. Additionally, the positions adopted by Italy and the other state, alongside Italian case law and administrative practice, must also be considered.

Moreover, the amendments to Article 5(4), introduced in response to BEPS challenges, lays an additional layer of complexities. These changes dilute the original intent of the "negative list", which aimed at simplification by providing a clear list of exempt activities. Instead, they require a thorough assessment of the nature of the listed activities and introduce anti-abuse provisions to prevent fragmentation of operations. While these amendments are necessary to address evolving economic realities, they may pose additional challenges for mega-projects, potentially hindering their successful implementation and realization of full potential.

Lessons from Leading Cases

Is an Hyperloop System A Megaproject? An Italian Case Study Between the Cities of Milan and Rome

Fabio Borghetti(✉), Federico Ursino, and Michela Vischi

Mobility and Transport Laboratory, Design Department, Politecnico di Milano, Milan, Italy
fabio.borghetti@polimi.it

Abstract. This paper analyzes the potential of implementing an Hyperloop transportation system between Milan and Rome in Italy, exploring its feasibility as a megaproject. The two cities, about 530 km apart, are home to some of the country's largest institutions and multinational companies, making them important generators and attractors of mobility demand . The Hyperloop system, introduced by Elon Musk in 2013, utilizes magnetic levitation and a vacuum tube to transport passengers and goods at speeds exceeding 1000 km/h. Given the scale and complexity of such a project, it can be classified as a megaproject, which involves significant capital investment, multiple stakeholders, and substantial socio-economic impacts. The study includes a cost analysis of both CAPital EXpenditure (CAPEX) estimated around € 20.5 billion and OPerational EXpenditure (OPEX) estimated around 187 million €/y (euro/year) and examines different revenue scenarios. The first scenario (Maximum revenue scenario) maximizes revenue for the infrastructure operator, the second (Maximum transport demand) maximizes demand and therefore the service offered, and the third (Elon Musk scenario) sets the ticket price at €20. Among the scenarios analyzed, it was found that the first profits from the Hyperloop would be generated starting from the 21st year, with a profit of €2.5 billion beginning in the 48th year, assuming maximum revenue. The results highlight that, despite the high initial costs, the system could provide long-term economic benefits, particularly through a public-private partnership model. The conclusion underscores the potential of the Hyperloop to revolutionize mobility between Milan and Rome, making it a competitive alternative to current transportation options. Future research could further explore cost optimization and sustainability improvements.

Keywords: Hyperloop · Mobility Project · Megaprojects · Transportation Planning · Sustainable Mobility · Transport modelling · Mobility · Transport Economics · High-speed transportation

1 Introduction

The Hyperloop transportation system was introduced by Elon Musk in 2013 [1]. This innovative system uses magnetic levitation on one side and a vacuum tube (tunnel) on the other to transport people and goods at high speeds (over 1000 km/h).

E. Favari et al. (Eds.): MeRIT 2024, LNCE 600, pp. 135–154, 2025.
https://doi.org/10.1007/978-3-031-92797-3_10

Magnetic levitation avoids physical contact between the capsule designs with aerodynamic shape (vehicle) and the track. The vacuum tube allows the capsules, made of lightweight materials, to move at a high speed while reducing air resistance.

Today, several studies and research are addressing the Hyperloop system, which in complexity can be considered a megaproject. But what then are megaprojects? Why do they play an important role in the development of our society today? What impact do megaprojects have in the mobility and transportation sector?

It is difficult to provide a single definition of megaprojects. Megaprojects can be defined as "large, complex initiatives that usually require an investment of at least $1 billion, years to develop and build, involve a multiplicity of public and private actors, have a transformative impact, and affect millions of people" [2, 3].

Considering the spatial issue, megaprojects significantly influence economic, social and environmental development, making sustainability a central theme. For example, Wang et al. (2020) propose a classification of megaproject sustainability into six categories:

- mega-event,
- socio-economic,
- growth coalition,
- environmental license,
- stress points,
- private sector.

In addition, the framework of "3Ps" (Purpose, People and Process) is proposed as a guide for the sustainable development of such projects [4].

Beyond the large capital investment, megaprojects are characterized by their extreme complexity, high risks, long duration, and significant impact on the community, economy, technological progress, and environment at the regional or national level. When considering the social aspect, they can be described as rapid and profound transformations of landscapes, requiring close cooperation between capital and state authorities [5].

Then there is an additional aspect concerning the social responsibility of megaprojects (MSRs) that is essential for their sustainable development. Lin et al. (2017) propose a holistic indicator system that considers the project life cycle's dynamic nature, stakeholders' heterogeneity, and the interaction between social responsibility and the project [6]. Some studies focus on case studies related to transportation infrastructure and mobility [7–9].

Gil and Pinto (2018) explore the issue of megaproject planning within polycentric organizations, highlighting how the developer retains decision-making authority over high-level choices but shares local decisions with autonomous stakeholders to better manage local disputes [10]. Gil and Fu (2022) and Gil (2023) address the issue of rising costs and delays by associating them with changes in project governance [11, 12].

It should be considered that trust, because of the many actors involved, is critical to the success of megaprojects. Ceric et al. (2021) identify six key aspects of trust in megaprojects [13]:

- trust and project success;

- institutional and public trust;
- trust in implementation methods;
- trust in contracting;
- trust as a governance mechanism;
- trust as competence of the project manager.

According to Gellert and Lynch (2003) [14], megaprojects are projects that rapidly, intentionally and profoundly transform landscapes, requiring the coordinated application of capital and state power. The authors divide megaprojects into four main categories:

- infrastructure (such as ports, railways, urban water and sewage systems);
- extraction (minerals, oil and gas);
- production (such as industrial plantations and export processing zones);
- consumption (large tourist facilities, shopping malls and theme parks).

More generally, the expansion of mega-projects globally is estimated to be in full swing, with an estimated annual market of $6 to $9 trillion [15].

The main goal of this work is to investigate the financial and economic feasibility and sustainability of a Hyperloop transportation system developing a framework usable in different contexts and presenting a real case study.

The paper is organized as follows: Sect. 2 presents the background regarding the different studies on the Hyperloop system, Sect. 3 describes the methodological approach used, Sect. 4 presents the Italian case study involving the connection between the cities of Milan and Rome, and Sect. 5 presents the conclusions and possible developments of the work.

2 Background

In the technical and scientific literature field, several authors have studied and continue to investigate the Hyperloop transportation system from multiple perspectives, including technological, organizational, and economic (Michelberger et al., 2022) [16]. The following summarizes the main studies that help frame the presented work.

In a general analysis, Mitropoulos et al. (2021) offer a review of the existing literature on the Hyperloop, considering various aspects such as the capsule, infrastructure, communication system, and identification of stakeholders involved [17]. Their analysis shows that research activities on the Hyperloop are distributed almost equally between Europe (39%) and Asia (38%), with a concentration of stakeholders in Spain (26%) and Germany (20%). Work focuses on the areas of capsule (37%) and tube (28%) development, with particular attention to impacts in terms of safety (35%), energy (33%) and cost (30%). However, the absence of full-scale structures makes system testing complex.

In the work of Planing et al., (2025) the issue of user acceptance of the hyperloop system is addressed. The study covers a sample of 387 participants in the Netherlands. On the one hand, performance expectations (e.g., high speed, comfort, environmental benefits) support acceptance of the hyperloop. On the other hand, safety concerns (such as e.g., technology failure, low-pressure environment) were identified as a rejection factor [18].

Premsagar and Kenworthy (2022) point out that there is still no unanimous consensus among scholars on the technical feasibility of the system [19]. Guerrieri (2022), on the other hand, compares innovative transportation systems-including Transrapid, Hyperloop, and HeliRail-with high-speed trains, concluding that the latter appears to be the better solution except in special cases [20].

Rajendran and Harper (2020) analyzed the feasibility of a Hyperloop link between Los Angeles and San Francisco, applying a DMADV (Define, Measure, Analyze, Design, Verify) approach to understand system operations. The results indicate that the number of capsules and the variability of commuter volume significantly affect system performance [21].

Voltes-Dorta and Becker (2018) examined the impact of the Hyperloop on the choice behaviors of airport users between San Francisco and Los Angeles, suggesting that introducing the system could alter airport competition [22].

On the technical side, Tudor and Paolone (2021) focus on the energy optimization of the system, analyzing aspects such as internal tube pressure and capsule design, and proposing adopting a magnetic levitation system to reduce consumption [23].

Van Goeverden et al. (2018) compare the operational, financial, and environmental performance of the Hyperloop with other transportation systems, showing that the Hyperloop has low capacity and high costs per passenger kilometer. However, it offers advantages regarding reduced energy consumption and minimal greenhouse gas emissions [24].

Kowal et al. (2022) analyze the possible implementation of the system in Poland, focusing on the estimated cost and demand for capsules [25]. Hansen (2020), on the other hand, analyzes the Hyperloop link between Germany and California, exploring various technical and feasibility issues [26].

Merchant and Chankov (2020) offer an analysis of Hyperloop networks in Europe, examining the importance of predefined routes and their influence on passenger mobility [27], while Gkoumas and Christou (2020) identify the safety and commercialization challenges of the system, providing a comprehensive overview of ongoing technological developments [28].

Also in the work of Badgujar et al., 2024, factors that can determine the success or failure of a Mumbai-Pune Hyperloop system in India are analyzed. One interesting issue concerns the stability of the government, which is a crucial factor in the project's success; any government change can affect the project's cost and feasibility. A survey assessing public confidence in the government's ability to manage the project responsibly revealed that 71.6% of respondents do not trust the government for a megaproject like Hyperloop. In addition, the survey indicates that the Hyperloop system may not be affordable for middle- and lower-class taxpayers in society; thus, it would be an unpopular option for most people [29].

Finally, studies such as those by Dudnikov (2017) and Werner et al. (2016) focus on using the system for freight transport, evaluating technical and economic parameters related to capsules and infrastructure. These studies provide methods for estimating a Hyperloop line's potential demand and costs, offering a theoretical and practical approach to its feasibility [30, 31].

This paper analyzes the Hyperloop system over long routes, comparing different transportation services and defining innovative performance levels for sizing the system. It combines various aspects of a feasibility study, investigating the opportunity of building a hyperloop line and considering technical, transportation and economical issues.

3 Methodology

The proposed methodology is divided into two main phases: transport demand analysis and economic feasibility analysis. This work focuses particularly on the second phase. Regarding the demand estimation, reference is made to a previous study by the authors, where SP - Stated Preference surveys [32] were conducted. The results of these surveys allowed for the estimation of user choice probability and a preliminary sizing of the transportation system. Based on the results obtained from the demand analysis, it was possible to analyze potential costs and revenues.

To analyze potential costs and revenues, the construction costs of the entire infrastructure (CAPEX - CAPital EXpenditure), operational costs (OPEX - OPerational EXpenditure), and revenues from ticket sales are evaluated.

As for CAPEX can initially be estimated by breaking down the system into its main components and assessing the theoretical cost of each. The same approach can be used for estimating operational costs (OPEX).

A scenario-based approach is suggested to better understand the potential revenues of the infrastructure and not limit the analysis to a single value. In this case, the adopted methodology starts from the use of the transport scenarios previously defined in the transportation analysis, where revenues are calculated by multiplying the probability of choosing the Hyperloop system by the number of annual passengers on the analyzed route and by the ticket price for that scenario, as illustrated in Eq. (1).

$$R = P^* \, NP^* \, T \tag{1}$$

Where:

- R = Hyperloop system revenues;
- P = probability of choosing the Hyperloop system instead of the other transportation systems;
- NP = number of passengers per year between an Origin-Destination pair;
- T = Hyperloop system ticket cost for an Origin-Destination pair.

To evaluate the economic feasibility and sustainability of the project in the long term, it is necessary to calculate the trends in costs, revenues, and profits (or potential losses) over defined period (usually around 50 years for infrastructure projects).

To do this, it is essential to introduce several elements that can simulate the variation in costs and revenues over time:

- Transport demand increases over time, and the share of Hyperloop usage is also likely to grow; consequently, an annual growth rate of demand can be introduced, directly impacting revenues.
- Similarly, inflation is considered, which will increase revenues on one hand and operational costs on the other.

- Finally, CAPEX will be amortized over a certain number of management years, with an interest rate on the capital that must be appropriately defined.

Given these elements, it is possible to calculate the annual cash flow of the transportation system.

4 Case Study: The Milan-Rome Connection in Italy

In the following chapter, an Italian case study concerning the potential construction of an hyperloop line between Rome and Milan, is analyzed, as showed in Fig. 1.

Fig. 1. Location of the cities of Milan and Rome and their relative distance.

The choice of this route was made by considering various factors. In addition to the high population density of the two metropolises (more than four million inhabitants), it should be noted that they are home to some of the largest institutions and multinational companies in the region, making them hubs that generate mobility demand. Introducing the Hyperloop would facilitate the displacement of people working between these two cities, allowing them to save time and thus increasing commuter demand.

Another aspect considered is the current number of passengers traveling this route yearly. In 2022, it is estimated that more than six million people reached the two cities through the various available modes of transportation. The demand is hypothesized to be distributed across the various modal alternatives as follows: high-speed train (AV) accounts for 80%, followed by air travel at 13%. In comparison, the option of traveling by car is chosen by only 7% of users.

The economic analysis presented in this work is based on traffic data obtained from the demand survey. It proposes a series of economic scenarios that show the potential revenues that can be generated. Fig. 2 shows a diagram of the different phases of the work.

Table 1 shows the attributes used in the demand analysis and the values proposed to the users (for each attribute) during the SP surveys.

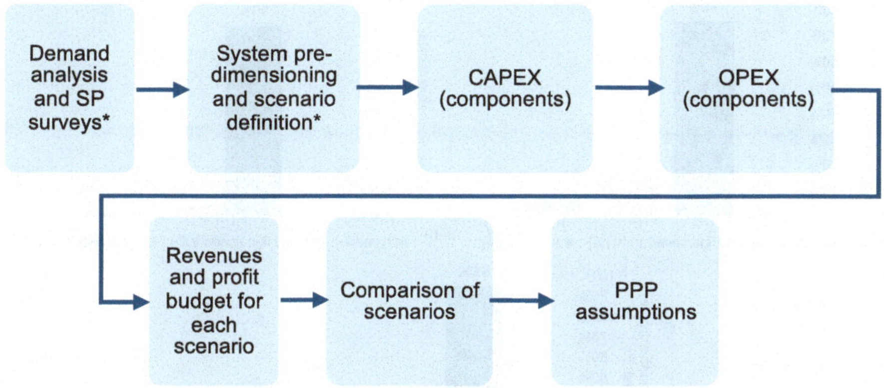

Fig. 2. Schematical view of the proposed framework *Issues explored in previous work by the authors [32].

Table 1. Modal distribution and attributes considered in the survey.

Modal alternatives	Cost [€]	Transport system frequency [min]	Travelling time [min]	Station entry and exit time [min]
Hyperloop	80; 150; 200	4; 8	30; 45; 60	15; 30; 45
High speed train	50; 100; 150	15	150; 180; 210	15; 30; 45
Airplane	60; 120; 180	60	45; 60; 80	45; 60; 90

Starting from the attributes listed above and using the Logit Model, the attributes were varied to generate three different scenarios [32]. The first scenario (Maximum Revenue Scenario) aims to maximize revenue for the infrastructure operator, with a ticket price of approximately €125. The second scenario (Maximum Transport Demand) maximizes demand with a ticket price of €80, thereby maximizing the service offered. Finally, the third scenario (Elon Musk Scenario) is inspired by Elon Musk's idea of setting the ticket price at €20. Below, in Fig. 3, the graphs showing the probabilities of choice for each modal alternative are presented.

Starting from these results, it is possible to develop the analysis of the potential costs and revenues. The main characteristics of the track and the transport system have been defined to estimate the costs. Table 2 shows the key geometric characteristics of the system that have a greatest impact on the costs:

For the tunnels, a length of approximately 110 km has been estimated, considering the increase from those presents in high-speed rail due to the high speeds that require larger radii of curvature, thus making integration into the territory more challenging. Regarding the more transport-related aspects, the assumptions are detailed in Table 3.

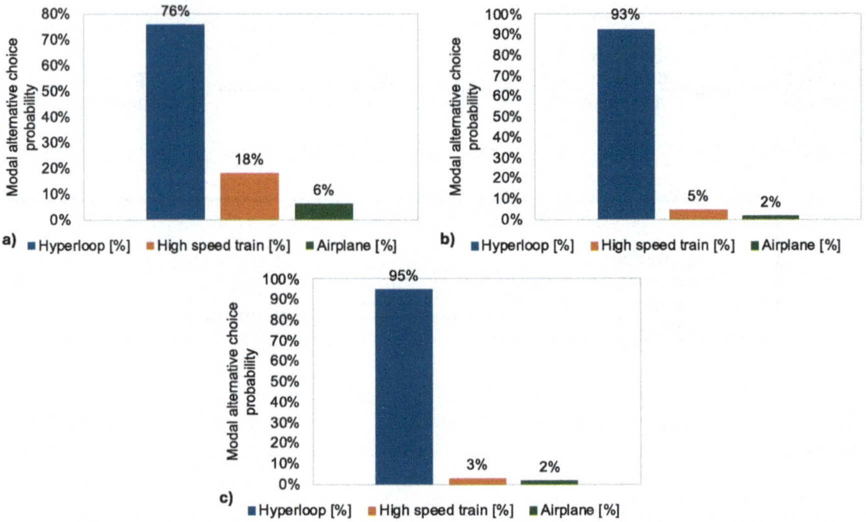

Fig. 3. Modal choice probability; a) Maximum revenue scenario, b) Maximum transport demand and c) Elon Musk scenario.

Table 2. Hypothetical track geometric characteristics.

Attribute	Value [km]
Track length	530
Tunnels length	110
Exchange length	6

Table 3. Transport system characteristics.

Attribute	Value [–]
Number of capsules	27
Number of stations	2

4.1 CAPEX

To evaluate the cost of constructing the infrastructure, reference is made to the unit costs proposed by Delft University [33]. Table 4 summarizes the main components analyzed and their respective costs per kilometer.

The following paragraphs provide an assessment of the unit costs of the various components of the project to detail the assumptions made to calculate the total cost of the project.

Table 4. Breakdown of CAPEX into unit cost elements.

Object	**Cost [€/km]**
Reinforced concrete pylons	225,000
Steel Tube	3,200,000
Inductrack	918,000
LSM Propulsion	12,800,000
Solar Panels	1,000,000
Vacuum pumps	45,000
Tunnels	5,400,000
Land expropriation	5,500,000
Other	1,200,000
Labor	5,000,000
Total	**35,288,000**
Object	**Cost [€]**
Safety Certification	202,000,000
Stations	1,400,000,000
Capsules	175,000,000
Total	**1,777,000,000**

4.1.1 Reinforced Concrete Pylons

The concrete towers are the support of the infrastructure. They should be sufficiently strong to bear the tube's weight and the capsules' passage (vehicles). Appropriate foundations must be built in the ground. The following assumptions have been made:

- The cost of the foundation is €750 per pylon (20 m x 0.2 m);
- Concrete costs €250/m^3, with 24 m^3 per pylon, the cost is €6000;
- The height is 5 m;
- The base area is 4 m^2;
- The distance between towers is 30 m.

4.1.2 Tube

The material proposed for the tube is steel. Therefore, the following assumptions have been made to calculate the amount of steel required and the related costs:

- External diameter of 2.7 m;
- Thickness of 25 mm;
- Density of 7860 kg/m^3;
- Price of €2/kg.

With these assumptions, approximately 800,000 kg of steel is needed for each kilometer of tube, with a cost of around €1,600,000. Considering two tubes, one for each direction, the total cost is approximately €3,200,000 per kilometer.

4.1.3 Inductrack

There is not available information regarding the cost of the Inductrack. The cost has been estimated by Delft University at €918,000 per kilometer, based on the cost of the aluminum required for its construction [33].

4.1.4 LSM Propulsion

Propulsion is necessary to generate the movement of the capsule. A Linear Synchronous Motor (LSM) is used for propulsion and braking. It is assumed that the motor will be installed along the entire tube length. The costs have been estimated as follows:

- Magnetic guidance: approximately €3,600,000 per kilometer;
- Communication systems: approximately €870,000 per kilometer;
- Power supply system: approximately €1000,000 per kilometer;
- Inverter: approximately €930,000 per kilometer.

Therefore, the total cost per direction is around €6,400,000 per kilometer; considering two tubes, the total cost is approximately €12,800,000 per kilometer.

4.1.5 Solar Panels

To power the system (at least partially), installing solar panels above the track is possible when it is not in a tunnel. The following assumptions have been made:

- Monocrystalline solar panels will be used;
- The cost is €200 per m^2 of panel;
- 5 m^2 of panel will be installed per meter.

Overall, the cost of the solar panels is €1000,000 per kilometer.

4.1.6 Vacuum Pumps

The pumps are used to reduce the pressure from 10,000 Pa to 100 Pa to minimize the aerodynamic resistance of the capsule. It is assumed that there will be 2 pumps per kilometer per direction, resulting in a cost of €45,000 per kilometer, assuming a cost of €22,500 per pump.

4.1.7 Tunnels

For the tunnel, the following assumptions have been made:

- Both tubes are placed within the same tunnel;
- The tunnel is made of reinforced concrete;
- The diameter is 8.5 m.

According to the British Tunnel Society, analyzing 21 tunnels in the European Union, the cost of a tunnel with a similar diameter range from £18 million to £39 million per kilometer [34]. The cost per kilometer primarily increases with diameter and decreases with the total length of the tunnel. The minimum cost for a tunnel with a diameter of 8.5 m is €26 million per kilometer, including all safety systems.

The total cost of the tunnels for the proposed route is €2,860,000,000, which translates to a cost for the entire route of approximately €5,400,000 per kilometer for the 530 km stretch.

4.1.8 Land Expropriation

The costs for design and land expropriation in an infrastructure project represent one of the highest cost items; in high-speed rail, these costs range from 5% to 10%. These costs include feasibility studies (technical and economic), technical drawings, land expropriation, and other additional costs such as legal and administrative fees, licenses, and permits.

On average, for high-speed rail projects, the total investment cost has been approximately €17.5 million per kilometer, with design and expropriation costs varying between about €two million and €five million per kilometer. Since the Hyperloop will also pass through densely populated areas, the maximum cost of approximately €5,500,000 per kilometer is considered.

4.1.9 Safety Certification Costs

To be operational, the Hyperloop needs to comply with specific safety standards. These standards correspond to certifications issued by specialized companies for existing transportation modes. These requirements should be met for both the infrastructure and the capsules. There are no specific safety regulations for the Hyperloop; therefore, the existing railway regulations are considered applicable. For railways, safety certification consists of four aspects:

- European regulation regarding interoperability;
- European regulation regarding safety;
- European regulation regarding maintenance;
- National regulations.

The costs related to the certification of a new train are in the range of several million euros. According to estimates from Delft University, it can be considered as 1% of the total infrastructure cost [33], which amounts to a cost of €202,000,000 for the route in question.

4.1.10 Stations

To estimate the cost of the stations, it is possible to refer to railway stations. The main dimensions of the station include:

- Area for elevators and platforms: approximately 6600 m^2;
- Transition area between platform and controlled pressure track: approximately 8000 m^2;
- Area dedicated to security: approximately 1000 m^2;
- Total: approximately 15,500 m^2.

Based on these assumptions, Delft University suggests comparing this station to the Berlin railway station, which cost approximately €700 million [33]. Although this cost may be overestimated, it is used to account for potential additional costs related to the Hyperloop. Considering that 2 stations for the Hyperloop have been proposed for the Rome-Milan route in this study, the total cost amounts to €1,400,000,000.

4.1.11 Other

In the previous paragraphs, the cost items of the system's main elements have been evaluated. This section analyses additional items that may significantly impact the project's final cost.

First, the systems that could be necessary have been defined. Following various research and comparisons, it has been deemed consistent with the work done so far to use the cost items proposed by Delft University, which have been validated in the literature [33]:

- Lighting: €10,000 per kilometer;
- Safety systems, such as emergency exits, ventilation, and first aid kits, estimated at €100,000 per kilometer;
- Depressurization units: €1000,000 per kilometer;
- Welding: estimated at approximately 1% of the total tube cost.

The total cost of these items for a bidirectional tube is €1,200,000 per kilometer.

4.1.12 Capsules

From a technological point of view, a Hyperloop capsule can be compared to a Maglev train. A magnetic levitation train with a capacity of 90 seats costs between €12.5 million and €15 million. Considering the need for greater structural resistance due to the pressure in the tubes, the higher value of approximately €15 million per capsule is adopted. A capsule with 40 seats would have a proportional cost, estimated at around €6.5 million. For the Rome-Milan route, a total of 27 capsules have been estimated, resulting in a total cost of €175,000,000.

4.1.13 Total Cost and Amortization

Considering the costs presented, the total cost of the Rome-Milan route in Hyperloop is €20,650,000,000. For a more precise economic analysis, it is assumed that the amount will be amortized over 50 years with a capital cost of 5%, using the formula provided in Eq. (2).

$$A = \frac{P^{*}r}{1-(1+r)^{-n}} \tag{2}$$

Where:

- A = annuity amortization constant;
- P = project total cost;
- r = cost of capital;
- n = number of years of amortization.

Consequently, the annual depreciation installment is approximately €1,132,000,000.

4.2 OPEX

For the operating costs, reference is made to the publication by Hyperloop Transportation Technologies in the Great Lakes Hyperloop Feasibility Study [35], where the following costs were analyzed, as shown in Fig. 4:

- Onboard staff
- Capsule maintenance
- Administration
- Track maintenance
- Insurance
- Pump maintenance
- Tube maintenance
- Energy for the pumps
- Stations
- Onboard services
- Energy for propulsion

The study proposes to evaluate the operational costs as 0.069 €/pkm (passengers per kilometer). Supposing to have five million passengers on a 530 km route it will result in approximately €187,000,000 per year in operating costs, as shown in Table 5.

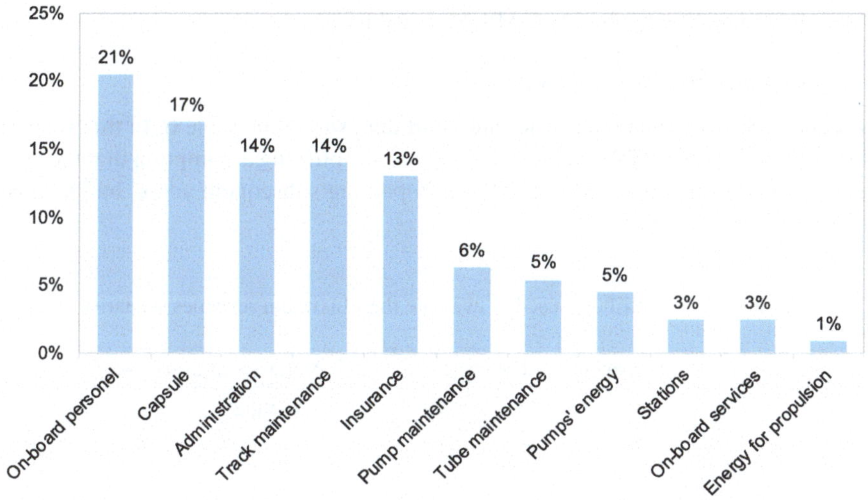

Fig. 4. Cost breakdown and weight regarding the total OPEX costs.

4.3 Revenues and Budget for Each Scenario

For each of the three scenarios presented at the beginning of the section (Maximum revenues, Maximum demand, and Elon Musk), annual revenues (or possibly losses) were calculated over a 50-year time horizon. To do this, as described in the methodology, the following assumptions were adopted:

Table 5. Operational costs.

Onboard staff	39,000,000 €
Capsule maintenance	31,000,000 €
Administration	26,000,000 €
Track maintenance	26,000,000 €
Insurance	24,000,000 €
Pumps maintenance	11,000,000 €
Tube maintenance	9,000,000 €
Energy for the pumps	9,000,000 €
Stations	5,000,000 €
Onboard services	5,000,000 €
Energy for the propulsion	2,000,000 €
Total	**187,000,000 €**

1. Transportation demand growth rate of 2%;
2. Inflation equal to 2%.
3. Cost of capital to be applied to CAPEXs equal to 5%.

4.3.1 Maximum Revenues Scenario

This scenario was implemented from the input data shown in Table 6. In this scenario, the first year with a profitable budget (making the simplifying assumption that the losses of previous years are smoothed out without impacting subsequent years' budgets) is at year 21, as shown in Fig. 5.

Table 6. Main attributes used to evaluate the Maximum revenues scenario.

Ticket price	Percentage of HP users	Numbers of users year 0
125€	76%	4,864,000

4.3.2 Maximum Demand Scenario

This scenario was developed from the input data shown in Table 7. As shown in Fig. 6, it is observed that the first year with a profitable budget (making the simplifying assumption that the losses of previous years are smoothed out without impacting the budget of subsequent years) is in the 26th year.

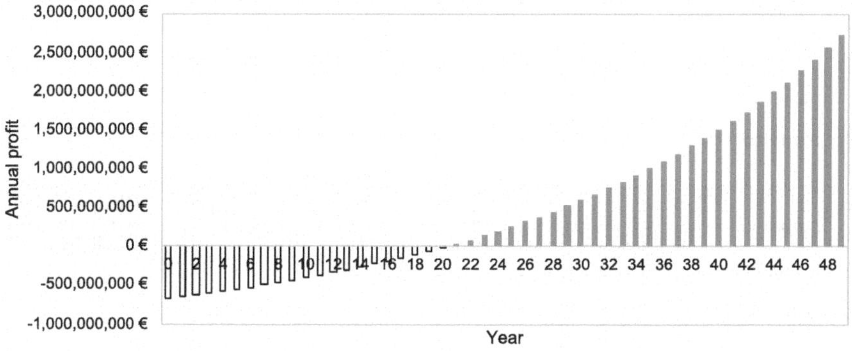

Fig. 5. Concessionaire's annual profit in the Maximum revenues scenario.

Table 7. Main attributes used to evaluate the Maximum demand scenario.

Ticket price	Percentage of HP users	Numbers of users year 0
80€	93%	5,952,000

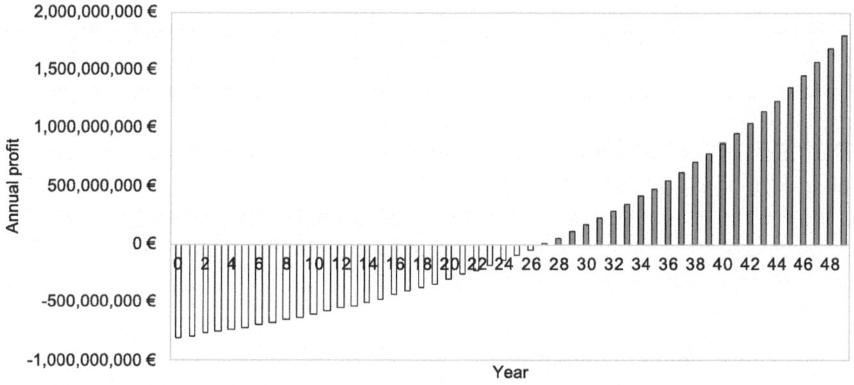

Fig. 6. Concessionaire's annual profit in the Maximum demand scenario.

4.3.3 Elon Musk Scenario

This third scenario was implemented from the assumption made by Elon Musk in 2013 as shown in Table 8. In this scenario, the losses would be substantial, and in the 50th year there would still be an annual loss of more than €600,000,000 as shown in Fig. 7.

Table 8. Main attributes used to evaluate the Elon Musk scenario.

Ticket price	Percentage of HP users	Numbers of users year 0
20€	95%	6,080,000

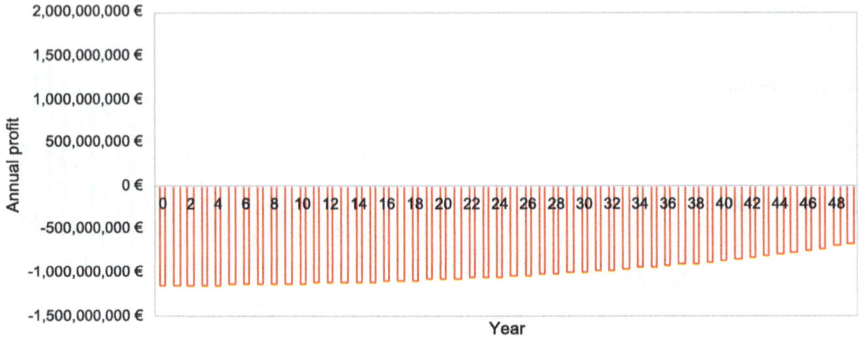

Fig. 7. Concessionaire's annual profit in the Elon Musk scenario.

4.3.4 Scenario Comparison

The "Maximum revenues" scenario is the most sustainable from an operator and investment perspective, maximizing revenues while attracting a share of transportation demand that is still high. The ticket price, however, might be prohibitive for many users, and thus, usage by business-type customers moving for business purposes might be assumed. On the other hand, the "Maximum demand" scenario would cause the Hyperloop to be chosen in almost all cases by having a price almost equal to that of its competitors. Still, the system's performance makes it preferred to the other modal alternatives proposed during the SP surveys (high-speed train, airplane, and automobile).

The "Elon Musk" scenario is not sustainable since, according to the data collected, the increase in transportation demand would be negligible. Comparing the probability of choosing the Hyperloop system between the maximum demand scenario and the "Elon Musk" scenario, the revenues for the operator would be 4 times higher in the former case by having to give up only 2% market share.

4.3.5 Private-Public Partnership

Finally, we tried to hypothesize the framework of a public-private partnership that would make the project economically viable. In such a hypothesis we would have two main actors:

1. The State (directly or indirectly through public companies) sponsoring the project;
2. A Special Purpose Vehicle (following SPV) composed of private companies that would build and operate the project for the next 50 years.

Focusing on the scenario called "Maximum revenues," the following assumptions are made:

1. The construction of the work takes place with 20% non-repayable public funding, with a total amount of €4,130,786,600;
2. Consequently, the amount to be amortized by the SPV turns out to be €16,523,146,400 with a constant annual installment of €905,084,019;

A graph of the SPV's annual operating performance trend is shown in Fig. 8.

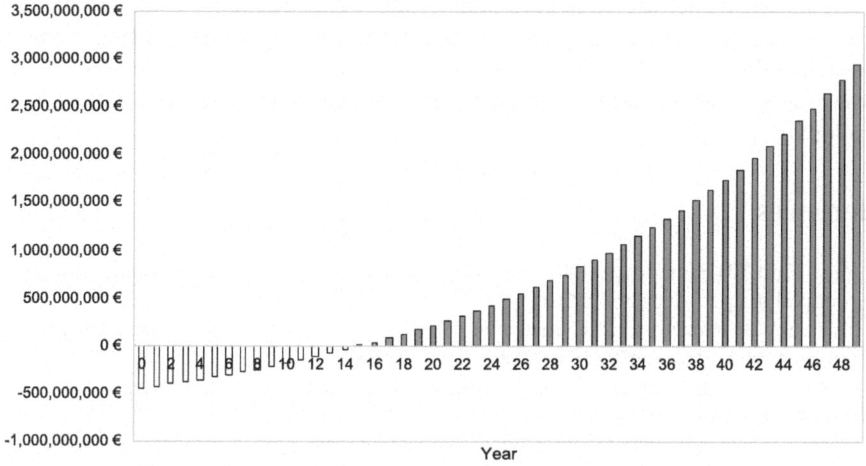

Fig. 8. Graph of the SPV's annual operating performance trend.

The budget appears negative for the first 14 years of the system's operation. Given the magnitude of the loss in the first few years, it is proposed that:

1. The State takes over to cover the losses for the first 14 years;
2. From the 15th year, the SPV will take over the operation of the project and pay the state a concession fee equal to 10% of revenues;

In this scenario, the SPV will have returned to the state the amount invested in the first 14 years to cover losses (net of capital revaluation).

5 Conclusions

The Hyperloop system presents significant performance potential, reaching up to 1200 km/h speeds, making it an innovative and interesting solution for connecting major cities like Rome and Milan. With reduced travel times, connecting Rome and Milan in about 30 minutes it could facilitate the emergence of new commuting dynamics between the two main Italian cities, regardless of the ticket price.

From an economic perspective, although construction costs (CAPEX) are particularly high, estimated at around €20.5 billion, the operational costs (OPEX), amounting to about €187 million, could be manageable compared to the potential annual profit could reach approximately €2.5 billion in the long term, making the system a competitive option in the long run.

Considering the scenarios examined in this paper, the public-private partnership solution appears to be the most sustainable economically. However, exploring additional cost scenarios in the future would be advisable to make the Hyperloop system even more sustainable and accessible. An in-depth analysis of these alternatives could improve this innovative infrastructure's economic effectiveness and adoption.

Nevertheless, the study still has some limitations that could be addressed in the future. Many costs were hypothesized, as there are currently few studies on the Hyperloop system, making it challenging to accurately estimate construction and operational costs. Therefore, exploring additional cost scenarios in the future would be prudent to enhance the sustainability and accessibility of the Hyperloop system. A thorough analysis of these alternatives could further improve this innovative infrastructure's economic effectiveness and adoption.

References

1. Musk E (2013) Hyperloop Alpha. https://www.tesla.com/sites/default/files/blog_images/hyperloop-alpha.pdf. Accessed 8 Oct 2024
2. Flyvbjerg B, Bruzelius N, Rothengatter W (2014) Megaprojects and risk: an anatomy of ambition. https://doi.org/10.1017/CBO9781107050891
3. Flyvbjerg B (2017) Introduction: the iron law of megaproject management. University of Oxford, Oxford
4. Wang T, Chan APC, He Q, Xu J (2022) Identifying the gaps in construction megaproject management research: a bibliographic analysis. Int J Constr Manag 22(9). https://doi.org/10.1080/15623599.2020.1735610
5. Söderlund J, Sankaran S, Biesenthal C (2017) The past and present of mega-projects. Proj Manag J 48(6). https://doi.org/10.1177/875697281704800602
6. Lin H, Zeng S, Ma H, Zeng R, Tam VWY (2017) An indicator system for evaluating megaproject social responsibility. Int J Proj Manag 35(7). https://doi.org/10.1016/j.ijproman.2017.04.009
7. Favari E (2020) Sustainability in (mega)project management—a business case for project sustainability. In: SpringerBriefs in applied sciences and technology. https://doi.org/10.1007/978-3-030-39354-0_2
8. Rizzi P, Turci L, Favari E (2023) Infrastructure and value creation for local systems: the case of Brescia-Padua A4. In: Lecture notes in civil engineering, 342 LNCE. https://doi.org/10.1007/978-3-031-30879-6_20
9. Caccialanza A, de Nito E, Canonico P, Favari E (2023) Mega-projects and social impact evaluation: the difficult (un)balanced inclusion of social needs in current practices. In: Lecture notes in civil engineering, 342 LNCE. https://doi.org/10.1007/978-3-031-30879-6_2
10. Gil N, Pinto JK (2018) Polycentric organizing and performance: a contingency model and evidence from megaproject planning in the UK. Res Policy 47(4). https://doi.org/10.1016/j.respol.2018.02.001
11. Gil N, Fu Y (2022) Megaproject performance, value creation, and value distribution: an organizational governance perspective. Acad Manage Discov 8(2). https://doi.org/10.5465/amd.2020.0029
12. Gil NA (2023) Cracking the megaproject puzzle: a stakeholder perspective? Int J Proj Manag 41(3). https://doi.org/10.1016/j.ijproman.2023.102455
13. Cerić A, Vukomanović M, Ivić I, Kolarić S (2021) Trust in megaprojects: a comprehensive literature review of research trends. Int J Proj Manag 39(4). https://doi.org/10.1016/j.ijproman.2020.10.007
14. Gellert PK, Lynch BD (2003) Mega-projects as displacements. Int Soc Sci J 55(175). https://doi.org/10.1111/1468-2451.5501002
15. Wang G, Wu P, Wu X, Zhang H, Guo Q, Cai Y (2020) Mapping global research on sustainability of megaproject management: a scientometric review. J Clean Prod 259. https://doi.org/10.1016/j.jclepro.2020.120831

16. Michelberger F, Plavec M, Pyrek K, Graf P (2022) The hyperloop concept development, possible applications and critical analysis. In: Marinov M, Piip J (eds) Sustainable Rail Transport 4. Lecture Notes in Mobility. Springer, Cham. https://doi.org/10.1007/978-3-030-82095-4_14
17. Mitropoulos L, Kortsari A, Koliatos A, Ayfantopoulou G (2021) The hyperloop system and stakeholders: a review and future directions. Sustainability 13(15). https://doi.org/10.3390/su13158430
18. Planing P, Hilser J, Aljovic A (2025) Acceptance of hyperloop: developing a model for hyperloop acceptance based on an empirical study in the Netherlands. Travel Behav Soc 38. https://doi.org/10.1016/j.tbs.2024.100887
19. Premsagar S, Kenworthy J (2022) A critical review of hyperloop (ultra-high speed rail) technology: urban and transport planning, technical, environ-mental, economic, and human considerations. Front Sustain Cities 4. https://doi.org/10.3389/frsc.2022.842245
20. Guerrieri M (2022) Hyperloop, HeliRail, transrapid and high-speed rail systems. technical characteristics and cost-benefit analyses. Res Transp Bus Manag 43. https://doi.org/10.1016/j.rtbm.2022.100824
21. Rajendran S, Harper A (2020) A simulation-based approach to provide in-sights on hyperloop network operations. Transp Res Interdiscip Perspect 4. https://doi.org/10.1016/j.trip.2020.100092
22. Dorta V, A., Becker, E. (2018) The potential short-term impact of a hyperloop service between San Francisco and Los Angeles on airport competition in California. Transp Policy 71:45–56. https://doi.org/10.1016/j.tranpol.2018.07.013
23. Tudor D, Paolone M (2021) Operational-driven optimal-design of a hyper-loop system. Transp Eng 5. https://doi.org/10.1016/j.treng.2021.100079
24. van Goeverden K, Milakis D, Janic M, Konings R (2018) Analysis and modelling of performances of the HL (hyperloop) transport system. Eur Transp Res Rev 10(2). https://doi.org/10.1186/s12544-018-0312-x
25. Kowal B, Ranosz R, Klodawski M, Jachimowski R, Piechna J (2022) Demand for passenger capsules for hyperloop high-speed transportation system-case study from Poland. IEEE Trans Transp Electr 8(1):565–589. https://doi.org/10.1109/TTE.2021.3120536
26. Hansen IA (2020) Hyperloop transport technology assessment and system analysis. Transp Plan Technol 43(8):803–820. https://doi.org/10.1080/03081060.2020.1828935
27. Merchant DV, Chankov SM (2020) Towards a European hyperloop network: an alternative to air and rail passenger travel. In: Paper presented at the IEEE International Conference on Industrial Engineering and Engineering Management, 2020 - December 128-132. https://doi.org/10.1109/IEEM45057.2020.9309907
28. Gkoumas K, Christou M (2020) A triple-helix approach for the assessment of hyperloop potential in Europe. Sustainability 12(19). https://doi.org/10.3390/SU12197868
29. Badgujar P, Sharma S, Rahul D, Sharma SK, Upadhyay RK (2024) Feasibility study of Mumbai-Pune Hyperloop corridor: challenges and imple-mentation. Res Transp Bus Manag 52. https://doi.org/10.1016/j.rtbm.2023.101089
30. Dudnikov EE (2017) Advantages of a new hyperloop transport technology. In: Paper presented at the Proceedings of 2017 10th International Conference Management of Large-Scale System Development, MLSD 2017. https://doi.org/10.1109/MLSD.2017.8109613
31. Werner M, Eissing K, Langton S (2016) Shared value potential of transporting cargo via hyperloop. Front Built Environ 2:17. https://doi.org/10.3389/fbuil.2016.00017
32. Borghetti F, Ursino F, Vischi M, Maja R (2023) Preliminary technical and economic analysis of a hyperloop line: case study from Italy. Eur Transp 90. https://doi.org/10.48295/ET.2023.90.7
33. Delft Hyperloop (2024) https://www.delfthyperloop.nl. Accessed 14 Oct 2024

34. Grose B, Hellier A (2018) Benchmarking tunnelling costs and production rates in the UK. Infrastructure and Projects Authority. https://www.gov.uk/government/publications/case-study-benchmarking-tunnelling-costs-and-production-rates-in-the-uk. Accessed 14 Oct 2024
35. Hyperloop TT, TEMS (2020) Great Lakes Hyperloop Feasibility Study by NOACA. https://www.hyperlooptt.com/hyperlooptt/wp-content/uploads/2022/11/Great-Lakes-Feasibility-Study-.pdf. Accessed 14 Oct 2024

Sustainability Disclosure and Legacy of Sports Mega-Events: The Case of Paris 2024 Olympic Games

Andrea Caccialanza[1], Ernesto De Nito[2(✉)], and Paolo Canonico[3]

[1] Università di Bologna, Bologna, Italy
[2] Università degli Studi di Salerno, Fisciano, Italy
edenito@unisa.it
[3] Università degli Studi di Napoli Federico II, Naples, Italy

Abstract. This paper investigates the management of mega-projects, using the Paris 2024 Olympic Games as a case study, and focuses on the evaluation of accountability, stakeholder engagement and legacy of this event. Employing a qualitative research approach, it analyses interim sustainability and legacy reports through document analysis, triangulating data from sustainability reports, official websites, and social media channels. The study adopts a framework classifying the tools and initiatives according to the dimension of *stakeholder recognition, stakeholder support,* and *stakeholder dialogue* to assess stakeholder engagement and project performance. While focusing on the pre-event, organizational, and event-start phases, it contributes to the literature on mega-project management by highlighting the challenges of a multi-dimensional disclosure, seek of legitimization and transparency, and limited stakeholder engagement, particularly the exclusion of dissenting voices from local communities. Our findings show the complexity of managing large-scale events and the need for integrated approaches to accountability of impacts, offering a preliminary evaluation of sustainability practices in a significant case-study within the context of temporary mega-sport-events. The study emphasizes the need for comprehensive post-event evaluations and provide a call for a clear understanding of the Games' long-term sustainability and legacy.

Keywords: Olympic Games · Mega-projects · Sport event · Sustainability · Legacy · Disclosure

1 Introduction

Mega-projects constitute a relevant area within project management discipline, involving large-scale and highly complex settings that significantly influence local communities and economies (Brookes and Locatelli, 2015). Defined by their enormous budgets, extended timelines, and potential to transform entire regions, mega-projects present unique management challenges (Vickerman, 2017; Wang et al., 2023). These major projects typically encompass the construction of major infrastructure - such as bridges,

E. Favari et al. (Eds.): MeRIT 2024, LNCE 600, pp. 155–174, 2025.
https://doi.org/10.1007/978-3-031-92797-3_11

airports, and urban developments - or the orchestration of globally significant events (Söderlund et al., 2017). This chapter contributes to the literature of the latter typology, specifically referring to the broader category of mega-projects (Sydow & Braun, 2018), and addressing the specific complexities of specific settings temporary in terms of execution, but with (potential) long-lasting effects and impacts (Flyvbjerg, 2014).

Due to the recognition of large scale-impacts at environmental, social and economic level, there has been also a growing awareness of the need for enhanced accountability in the execution of mega-projects (Landis, 2022). Since public funds are frequently allocated to fund these large scale-initiatives, stakeholders - including governments, taxpayers, and private investors - are increasingly demanding stringent performance evaluations and transparent reporting practices (Corazza et al., 2022). Therefore, this emphasis on accountability is crucial not only for mitigating inherent risks but also for ensuring that mega-projects achieve their intended objectives and deliver anticipated benefits (Dyer, 2017). Moreover, in an era in which environmental and social sustainability are of increasing importance, the discipline can contribute to ensure these projects are economically sound, environmentally responsible, and socially equitable (Caccialanza et al., 2022; Corazza et al., 2022).

A specific domain within the mega-projects field that deserves special attention is the organization of large sport-events (Canonico et al., 2024). Such events are one of the best-known examples of mega-projects, due to their grand scale, substantial budgets, and significant societal impacts (Pitsis et al., 2018). In fact, sport events introduce unique complexities, particularly from a financial standpoint, involving aspects like sponsorship management and the funding of both permanent and temporary infrastructure (Varrel and Kennedy, 2011). Therefore, what clearly emerges is that the impacts of temporary sport mega-events are not limited (in time and space) to the preparation and execution of the event, but they produce more permanent legacies on the local community (Al Hallaq et al. 2021; Babaei et al., 2023). Consequently, at least an estimation of the magnitude of those impacts must be addressed by the organizing committee of such events.

The literature focusing on sports events as mega-projects provides valuable insights into the practical application of project management principles, accountability measures, and sustainability practices (Rook and Heerdt, 2023). Investigating sports events within this context can enhance the debate within the broader discipline of mega-project management. This novel approach adds an incremental perspective to existing literature, contrasting with the traditional focus on mega-projects themselves (Cantarelli and Genovese, 2021). The emergence of this diverse and fragmented body of research inspires and justifies the effort to discuss a significant case study on the topic, presenting the preliminary results of the Paris 2024 Olympic games.

Paris 2024 Olympic Games seem to be expected to be a fresh example of cost-overruns-linked events (Budzier & Flyvbjerg, 2024). A recent study highlighted that the costs for the Paris Games have exceeded their original estimates by 115% (Budzier & Flyvbjerg, 2024). This is consistent with the trend of financial overruns seen in previous Games. Additionally, there are growing concerns about the long-term legacy of the Paris Games, as the reuse of infrastructure seems to be falling short of initial promises (Bourdin et al., 2024). These challenges reflect the broader difficulty of controlling and delivering long-term benefits.

Our research takes into account the assessment of impacts and legacies referred to this mega-event. According to recent studies, two main difficulties arise: the size of the urban population and the challenge of leaving behind a meaningful legacy (Bertolini and Sargolini 2023). Paris is an example of complex setting for a mega event since has to manage its vast population, within the largest urbanized areas in Europe, while also ensuring that the Olympic infrastructure serves the city's needs after the Games. The ability of the organizers to navigate these complexities will be a critical test of whether the Olympic model can evolve to become more sustainable and efficient.

This paper is organized as follows: the subsequent sections details the background and the methodology employed; the Sect. 4 presents and discusses the results obtained; finally, the paper concludes with reflections and a proposed research agenda.

2 Background

2.1 Sustainability of Mega-projects and Stakeholders' Engagement

Sustainability refers to the idea of enlarging traditional practices of mega project management to encompass economic, social and environmental impacts (Ahmed et al. 2022; Caccialanza et al. 2023). There are several dimensions in sustainable mega project management: the triple bottom line of environmental, economic and social aspects, the temporal and spatial life-cycle dimension, the stakeholder dimension and the values dimension (Wang et al., 2023). In our paper we will focus specifically on the stakeholder dimension.

Stakeholders are usually defined as entities that are affected by and can impact a project. Megaprojects lead to a large number of stakeholders with diverse interests and broad scope (Flyvbjerg, 2014). Because these projects heavily rely on integrating different kind of resources and meeting different project goals, managing stakeholders effectively is crucial. Proper stakeholder management is vital for harnessing the substantial social resources needed for successful project execution and for addressing the complex conflicts of interest that arise. Thus, effective stakeholder management is a key factor in megaprojects, aiming to enhance the management process and maximize overall benefits—economic, social, and environmental (Wang et al. 2020).

Due to the significant dependence on integrating social resources and the diverse requirements of project objectives, effective stakeholder management is far more important in megaprojects than in standard projects (Wang et al., 2023). Managing stakeholders effectively is crucial for leveraging the substantial social resources necessary for project execution and for navigating the complex conflicts of interest that may arise. As a result, stakeholder management is a fundamental aspect of megaprojects aimed at optimizing the management process and maximizing benefits. Currently, research on stakeholder management in megaprojects primarily concentrates on engaging stakeholders, analysing their needs, and developing management strategies (Xue et al. 2020). In particular, engaging stakeholders in megaprojects involves prioritizing the various demands of stakeholders. Particularly in the context of the mega sport events, issues related to the management strategies of different stakeholders are of paramount importance.

2.2 Prior Evaluation of Olympic Games Editions

The Olympic Games, as one of the world's most prestigious sporting events, offer opportunities for global visibility, national pride, and the development of infrastructure. However, these positive outcomes often come with significant costs and criticalities. These challenges range from massive cost overruns to political tensions and complex organizational structures that compromise efficiency and lead to problems with refers to stakeholders engagement and endorsement (Flyvbjerg 2017). In particular, this phenomenon is driven by cost overruns, political tensions, and organizational inefficiencies. One of the key issues related to the Olympic Games is the consistent and significant cost overruns, which have been documented across many editions of the Games. A study by Flyvbjerg et al. (2021) demonstrates that the Olympic Games hold the record for the highest average cost overrun of any type of megaproject, reaching 172% in real terms. This statistic, albeit susceptible of reductions, highlights the financial risks associated with hosting the Games, particularly as cities and countries often underestimate costs and overestimate potential benefits during the bidding process.

Lovallo et al. (2023) further explain this phenomenon by discussing the systemic biases present in the planning and bidding processes for the Olympic Games. They argue that political and organizational forces push planners to distort or misstate information, primarily for personal or institutional interests. As cities and nations compete globally to host the Games, applicants are incentivized to understate the likely costs of the event while simultaneously overestimating the long-term benefits, such as tourism and infrastructure development. This practice, coupled with the inevitable complexity of organizing such a large-scale event, leads also to significant financial overruns.

The generative mechanism for the Olympic cost overrun has been identified as driven by six causal factors: irreversibility, fixed deadlines, the "Blank Check Syndrome," tight coupling of processes, long planning horizons, and the "Eternal Beginner Syndrome" (Flyvbjerg et al. 2021). The fixed deadlines imposed by the International Olympic Committee (IOC) often force host cities into a corner, where they must complete infrastructure projects regardless of the rising costs. Similarly, irreversibility plays a crucial role, as projects cannot easily be paused or cancelled once initiated. These factors combine to create a scenario where financial overruns become almost inevitable and large part of prior literature based their analysis on the economic sustainability aspect of those events.

In addition to financial overruns, there are often discrepancies between the promises made by host cities during the bidding process and the actual outcomes following the Games (Chappelet & Junod 2006). The legacy of the Olympic Games is frequently portrayed as a long-term benefit, with improvements in infrastructure, tourism, and local economies. However, the experience of London 2012 reveals the "false promises" often made about the legacy (Flyvbjerg et al., 2016). London's bid promised that the Olympic infrastructure would be reused and repurposed, particularly in East London, to support future urban development. Yet, in practice, much of the infrastructure has been underutilized or has failed to deliver on these promises.

Political and organizational tensions further complicate the process of organizing the Olympic Games. These tensions exist not only between the host city and international organizations, such as the IOC, but also within the national and local governments responsible for the event. A study by Lovallo et al. (2023) emphasizes how political and organizational forces contribute to inefficiencies, as planners deliberately distort information to serve their interests. These forces can lead to tension between the different stakeholders involved in the event's planning and execution.

The need to balance the successful visibility of the Games with controlling costs and delivering the promised legacy further complicates the situation. Organizing the Olympics requires context-specific coordination between various political entities, including local governments, national administrations, and international organizations. This coordination is often fraught with tensions as different stakeholders prioritize different outcomes, further exacerbating the challenges of staying within budget and delivering on the promises made during the bidding process. Therefore, these false promises can lead to political tensions, as local and national governments are pressured to justify the massive expenses incurred by hosting the Games. Furthermore, the focus on creating a positive Olympic legacy often leads to a biased allocation of resources, favouring the Games' immediate needs rather than long-term urban development or local community support. This tension was evident in London and continues to pose challenges for future hosts, including Paris 2024.

3 Methodology

This study employs a qualitative research approach (Barrat et al., 2011) to investigate key factors that enhance and measure performance within sustainability-focused ecosystems (DiVito & Ingen-Housz, 2021). Specifically, we utilize an exploratory single-case research design to examine successful implementations, similar to methods used by previous researchers when analysing best practices or worst scenarios (Eisenhardt, 1989; Yin, 2018). In literature, the case study approach is particularly well-suited to abductive reasoning, enabling the integration of general theoretical frameworks with the specific contextual details that emerge from individual cases (Ketokivi & Choi, 2014). In fact, the case study methodology enhances the comprehension of specific concepts, such as narrative instruments – i.e., sustainability reporting - and processes – i.e., (disclosure on the) reporting process -, focusing on their application within the particular case rather than aiming for broad generalization (Stake, 2005).

Additionally, this case-based methodology lays the groundwork for a more systematic and comprehensive approach, which in later stages can lead to more grounded, robust, and generalizable outcomes compared to isolated case studies as mentioned in the research agenda (Yin, 2018). On the operational side, the iterative process of alternation of theoretical background and data from the sustainability reports allowed us to ground the data within existing literature while extending it to enhance generalizability and refine construct definitions, following the approach recommended by Eisenhardt (1989).

In this article, the document analysis framework outlined by Bowen (2009) is applied to the case study of the Paris 2024 Olympic Games. This framework is used to explore the role of various 'external' documents in understanding the stakeholders engagement and, more broadly speaking, sustainability efforts associated with the event.

The data were organized according to three phases of mega-events, following an adapted approach that focuses on the pre-event, organizational, and event-start phases, rather than the pre-, during-, and post-event phases proposed by Salama (2021). This adjustment was made due to the lack of available documents assessing the accountability of post-event legacy evaluations. For each phase, the specific engagement tools or initiatives for stakeholder engagement were identified and aligned with the engagement scope and the relevant stakeholder or stakeholder groups involved. Furthermore, future studies may apply similar or more extensive data collections on the phasis proposed by Salama (2021), as other authors such as Moggi et al. (2024) have already performed for the case of Expo 2020 Dubai. Similarly to the analysis of Moggi et al. (2024), we adopted the classification of Aksoy et al. (2022). In particular they identify three dimensions of stakeholder engagement that help manage various tools used simultaneously. The first is *stakeholder recognition*, where organizations acknowledge stakeholders' rights, feelings, and traditions, involving low engagement and primarily one-way communication. The second is *stakeholder support*, where organizations provide resources such as money or services, fostering moderate involvement through a one-way relationship. The final dimension is *stakeholder dialogue*, characterized by high engagement through two-way communication, allowing the exchange of information and ideas.

By analysing the type of documents utilized by the Paris 2024 organizers, and by assessing their benefits and limitations, this study highlights how document analysis can deepen our understanding of the Olympic Games' operations within the broader context of stakeholders' engagement and sustainable development process.

3.1 Identification, Assessment and Valuation of Paris 2024 Impacts

The interim sustainability and legacy reports for the Paris 2024 Olympic Games raise important questions about accountability, transparency, and stakeholder engagement, issues often discussed in project management. Crawford and Bryce (2003) already highlighted that while strict reporting standards are crucial for diverse stakeholders, commonly used methods like the logical framework can be insufficient for ongoing monitoring. This indicates that the Paris 2024 reports may face similar difficulties in capturing the full, long-term effects of their sustainability initiatives. Accordingly, Bruzelius et al. (2002) underscore the need for transparency and performance specifications to protect the public interest. In addition, public access to data and external reviews by experts are vital. Brunet (2021) discusses the inherent tensions between accountability, legitimacy, and efficiency in project management, tensions likely reflected in the Paris 2024 reports. Different stakeholder groups often have competing priorities, since institutions may focus on legitimizing decision-making, while other groups may demand greater accountability for broader impacts. This creates a challenge in balancing transparency with operational efficiency, as more transparency could reduce efficiency (Brunet, 2021).

The interim report on the Paris 2024 Legacy & Sustainability strategy provided an ongoing evaluation of the programs and actions implemented by Paris 2024 and its stakeholders. Since the event had already taken place, the report focused on the outputs and initial outcomes, offering a preliminary assessment of Pillar 2 of the strategy. According to Fig. 1, the evaluation followed a step-by-step approach based on the theory of change,[1] comparing the targeted outcomes with the results achieved during and after the event.

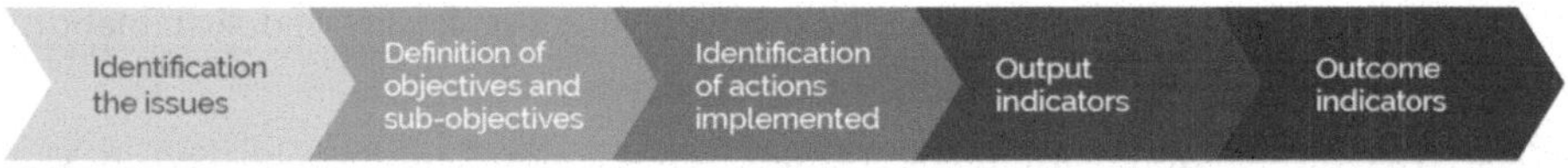

Fig. 1. Presentation of the stages in the evaluation process. Source: Interim Sustainability & Legacy Report, 2023, p. 31.

Moreover, Del Cerro Santamaria (2019) and Too and Weaver (2014) stress the importance of integrated communication systems, especially during early project phases. It is worth questioning whether Paris 2024's interim reports have sufficiently addressed stakeholder involvement during the planning stages or if they are more focused on ex-post evaluations, potentially limiting the scope of accountability. The second report, through the Fig. 2, also declares to be aimed to reinforce the confidence in the contribution of these actions to broader systemic changes, influenced by various factors. The analysis presented in the interim report was a work-in-progress evaluation, with more comprehensive reports scheduled for 2025 and 2029 to provide (short and long terms) ex-post evaluations and a more detailed understanding of the long-term impacts. More information about the documental analysis performed is discussed in the following sub-section.

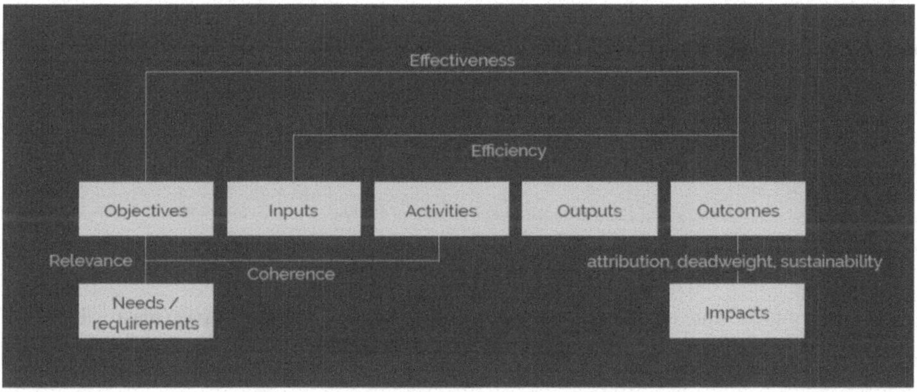

Fig. 2. Illustration of the impact value chain. Source: Interim Sustainability & Legacy Report, 2023, p. 32.

3.2 Documental Analysis and Data Sources

As mentioned in the first part of the methodological section, this study is based on a desk analysis of results using three data sources to ensure triangulation of information

[1] For more information about the adopted approach available at:- https://unsdg.un.org/sites/default/files/UNDG-UNDAF-Companion-Pieces-7-Theory-of-Change.pdf.

(Modell, 2005), specifically: sustainability reports, the official website, and institutional social media channels.

According to the official website of Paris 2024 Olympic games,[2] different reports presented the progressive actions undertaken with refers to legacy and sustainability reporting processed and impact evaluation:

- The "Delivering More Sustainable Games" Pre-Games Report, published in May 2024
- The interim evaluation report "Building the Social and Environmental Legacy of the Paris 2024 Games", published in November 2023
- The previous version of the Sustainability and Legacy Report published in August 2021

Furthermore, various studies offered useful frameworks for evaluating the contribution of major events to ecological transition, and these have been consulted to support the impact analysis in this report (see for other evaluations of projects' outcomes Müller et al. (2022)). Notable references[3] include the IOC Agenda 2020 + 5, ASOIF methodologies, the Commonwealth Secretariat guidelines, OECD recommendations, and the practical guide from the French Ministry for Ecological Transition. The reports' analysis of Paris 2024's ecological transformation mentioned as well to have incorporated insights from these frameworks, in order to allow higher comparisons and a well-rounded evaluation of its impact (Fondazione Milano Cortina 2026 2023, p. 37).

In the next sections are presented the preliminary results and descriptive evidences emerged from the documental analysis, while in the discussion section a broader comment on the comparison of those results with prior studies and further literature is proposed.

[2] All the mentioned reports are available at the following link of the official website:

- https://olympics.com/en/paris-2024/our-commitments/assessing-our-impact/sustainability-legacy-report.

[3] All the mentioned methodologies are available at the following links of the official websites:

- https://olympics.com/ioc/olympic-agenda-2020-plus-5.
- https://www.asoif.com/sites/default/files/download/asoif_common_indicators_for_measuring_the_impact_of_events.pdf.
- https://thecommonwealth.org/our-work/measuring-contribution-sport-sustainable-development-goals.
- https://www.oecd.org/fr/regional/global-events-recommendation.htm.
- https://www.ecologie.gouv.fr/sites/default/files/190729_CDDEP_guide_changement_de_comportement.pdf (French only).

4 Results and Discussion

The Paris 2024 Olympic and Paralympic Games have been planned with a strong emphasis on sustainability and environmental responsibility, showing a commitment to reuse, recycle, and repurpose temporary venues and structures. This is reflected significantly in the emphasis underlined on the relevance of this focus at (external) communication level. As declared in the institutional website, the organizing committee has committed to using approximately 95% of pre-existing or temporary venues for the event. Following the Games, all temporary structures and materials will be dismantled and either reused, repurposed, or recycled, fulfilling a pledge made during the bidding process. The operators responsible for these temporary installations are required to meet this commitment.

A key legacy of the Paris 2024 Games will be the transformation of the Olympic and Paralympic Village. Situated on a former industrial site that spans the areas of Saint-Denis, Saint-Ouen-sur-Seine, and l'île-Saint-Denis, the village development was already planned before the Games and has been accelerated due to the event. By 2025, the site will be redeveloped into a sustainable urban district, providing housing for nearly 12,000 residents, along with office spaces and retail shops. Sustainable construction techniques, including low-carbon materials, triple glazing, cooling floors, green spaces, and geothermal energy, are being incorporated to ensure the area is adapted to future climate conditions. During the Games, modular partitions have been used to accommodate 15,000 athletes, and all furniture will be chosen with reuse in mind, ensuring a sustainable second life for these items after the event.

In terms of reducing the environmental impact, Paris 2024 is focused on significantly lowering its carbon footprint. While the concept of "carbon neutrality" is typically applied at a global or regional scale, Paris 2024 is committed to reducing event-related carbon emissions by 50%. This reduction will be achieved by implementing sustainable practices across all aspects of event organization. For emissions that cannot be avoided, the organizing committee has committed to financing environmental projects aimed at combating climate change. These projects include initiatives in reforestation, forest preservation, and the development of renewable energy, all of which also provide benefits for local communities and biodiversity.

Since 2022, following recommendations from the French Environment and Energy Management Agency (ADEME), Paris 2024 has moved away from using the term "carbon-neutral" to describe the Games. This change reflects an acknowledgment that achieving full carbon neutrality through a single event is unrealistic. However, this shift in terminology does not imply a reduction in the Games' environmental ambitions. Instead, Paris 2024 remains committed to making substantial contributions to global efforts to reduce carbon emissions by lowering the event's impact and supporting sustainable projects that mitigate climate change.

Consequently, the Paris 2024 Olympic and Paralympic Games aim to leave a lasting legacy, not only in terms of sports infrastructure but also as a model for sustainable urban development and environmental responsibility. Through careful planning, innovative design, and a focus on reducing and offsetting carbon emissions, the Games are announced and disclosed to be positioned and set a new benchmark for environmentally conscious mega-events.

On the other hand, there is the consciousness that the path for a real engagement and endorsement of those initiatives, not only by official partners and supporters, but also local communities, is long and full of obstacles.

In a time-perspective, it is analysed the evolution of that effort and commitment in order to test if and how the announced engagement and expected endorsement of the different categories of stakeholders took place.

In the first report published in August 2021, this attempt is clearly stated, since one of the three pillars of the Paris' 2024 vision is exactly the 'engagement', namely "Starting today, Paris 2024 is rallying everyone and every community in France to its side: the sports movement, local and regional authorities and other public-sector stakeholders, the general public, corporate partners, etc." (Paris 2024, p. 5). With a more granular view, we propose for the pre-event period (2021) the classification of stakeholders' engagement initiatives according to the model of Aksoy et al. (2022) in Table 1.

One of the first evidences that emerged is the composition of the internal categories of stakeholders, employees. In 2020 the employed staff was composed by 379 employees, of which 291 with permanent contract. The gender equality quotas were ensured by the presence of 51% of woman and 5.01% of people affected by disabilities. On the other hand, no further details are provided regarding the gender distribution in relation to contractual arrangements or positions within the organizational structure since the narrative disclosure is focused on the on-boarding path and the development of significant skills/competences by the (managerial) team (see first sustainability and legacy report, 2021). Lately, at the starting point of the games in 2024, they reported 2059 employee, with half of them on permanent contracts, 52% women, and 3.84% with disabilities.

Appreciable is the reference to the identification of roadmaps set by the individual committees and working groups, although in the vast majority of cases there is a lack of a synoptic view of quantitative indicators related to measuring the expected performance by the organizing committee. The reference is inspectable through callouts and links to external documents outside the report, which limits its accessibility—an essential criterion in evaluating information in this type of report (p. 149). Similarly, the adoption of the GRI-2016 framework is commendable to ensure greater comparability of the disclosed performance, although it is limited to the "Core option" and lacks external assurance (p. 142). This choice limited the expected verifiability of the disclosed information, exposing the disclosure to proof of legitimacy issues and on-field-verification of the claimed or planned interventions and (positive) impacts. The concern is further fuelled when such large budgets, 3.9 billion € according to the 2021 report (p. 32) and 4.4 billion € according to the 2024 one (p. 38), are accompanied by control and disclosure tools that might be partially affected by a potential (undesired and unseen) bias. This bias is understood both as a subjective interpretation (internal and institutional) and as an analysis of only part of the estimated impacts, rather than a comprehensive assessment of all material issues in the broader sense, related to the more appropriate and critical concept of double materiality. In particular, no a-priori destination of the €52 million for Environmental Excellence and €50 million for Impact and Legacy are exposed, but just announced (see p. 32). The link worth the economic and/or financial dimension is not secondary since from the interdependences and integration of financial and non-financial management control systems and strategies, may emerge differences

Table 1. The level of stakeholder engagement - Pre-event (August 2021).

Dimensions of stakeholder engagement	Engagement tools or initiatives and involved stakeholders (IS)	Aim and description of the engagement tool
Stakeholder recognition	2 conferences, 3 webinars and 2 workshops by videoconference. IS: People with disabilities (PWD)	The programme includes training for managers and staff, communication to target audiences, adapted equipment. This programme reached 650 people
	Equality and Diversity Committee. IS: Minorities.	This committee draws up a shared roadmap to make sure the Paris 2024 Games promote equality and tackle discrimination
	ESS (SSE) 2024 Strategy Committee. IS: Contractors, specifically social enterprises	This committee monitors progress in the ESS 2024 Social and Solidarity Economy programme created to enable social enterprises to access Games contracts
Stakeholder support	Social Charter Monitoring Committee. IS: Institutional stakeholders and professionals	It ensures that Paris 2024 abides by the Social Charter it has signed
	Group For a Fit France. IS: Stakeholders in general	This group identifies concrete action to tackle sedentary lifestyles and encourage people to exercise
	Horizon 2024 Committee. IS: Schools and children	It promotes sports and exercise in schools, across France
Stakeholder dialogue	Legacy and Sustainability Steering Committee. IS: Stakeholders in general	Forum for the Organising Committee and its stakeholders to discuss their social and environmental endeavours
	Paralympic Games Legacy Committee. IS: People with disabilities (PWD) and Paralympic athletes	Meets to discuss and prepare a shared roadmap for universal accessibility at the Games, and to develop participation in sport and social and professional inclusion for people with disabilities

Sources: Institutional website and social media channels, and first sustainability and legacy report (2021).

between symbolics or substantial accountability processes. Those evidences have been yet largely discussed in literature, from the pioneristic studies of Leong (1991) to the most recent ones of scholars such as Millar et al. (2024).

The doubts further emerge when in the second edition of the interim-report (November 2023), the Chair of the Supervisory Committee claims "*The supervising committee is in agreement that the final 17 indicators selected are capable of demonstrating to a reasonable extent the social and economic legacy and impact of the Paris Games, if not the full breadth of impacts such an event will have*" (p. 5).

According to that report the 'Legacy & Sustainability strategy' is based on two pillars (p. 24): Delivering More Responsible Games; and Building the Social and Environmental Legacy of the Paris 2024 Games.

Furthermore, in July 2023, the Economic Co-operation and Development (OECD) released two guides for cultural, sporting, and business event organizers. These guides, created with input from academics, experts, and organizations like the Olympic Committee (IOC) and World Health Organisation (WHO), focus on evaluation approaches and reference indicators. They aim to help stakeholders measure the economic, social, and environmental impacts of major events. Paris 2024 have been the first to apply these OECD recommendations, following a 2019 IOC-OECD collaboration agreement.[4]

To further engage in the selective approach of indicators and assessment of (sustainability related) impacts, emerge a methodological and explanatory note (p. 30). Accordingly, the impact indicators in the guides had represented a selection of outputs and outcomes from the Legacy & Sustainability strategy, chosen to align with the OECD's dashboard guides, not the full list of them and with not consistent reference to that. Consequently, discrepancies had sometimes appeared between the titles of the indicators and the related outcomes. Secondly, the dashboard's economic and environmental indicators had been set to be detailed in a report on the Games' delivery, which was scheduled for publication in the second quarter of 2024. Albeit a selective decision of indicators could be made, upon proper not-reported justification, the second one is not aligned with the principles of optimal integration of financial and non-financial indicators. This happens even if are explicitly stated the issued and concerns about the quality evaluation of the reports according to the criteria of relevance and coherence (Interim Sustainability & Legacy Report, 2023, p. 34).

In addition, the analysis conducted by Ipama and Sport 1.5[5] in the interim report was based on a qualitative survey involving 31 individuals from 20 organizations and a quantitative survey of the Organising Committee's staff, with 156 respondents out of a total of 1559 (a 10% response rate). In our opinion, there might be doubts about the consistency of the findings due to the relatively low response rate from the workforce, which may affect the representativeness of the results. On the other hand, we also mapped,

[4] For more information:

- https://olympics.com/ioc/news/oecd-issues-independent-guides-to-measure-long-term-impacts-of-global-events-such-as-the-olympic-andparalympic-games.
- https://olympics.com/ioc/news/ioc-joins-forces-with-oecd-under-new-agreement.

[5] Ibama and Sport 1.5 are two consultancy firms.

as for the prior Table 1, the additional engagement initiatives in the interim report in Table 2. On the other hand, we are signalling that this synthesis is limited to the report of the actions undertaken solely by the organizing committee, not those in combination of, or conducted by, other partners. In particular, with refers to the legacy aspects of the educational initiatives, it emerges as albeit the involvement in the use of the educational platform seems relatively high in the interim-phase, a shortening in the use of those materials seems to be projected after the games. This may limit the evaluation of the legacy to shorter terms that the initial expectations.

Similarly, we conducted an analysis also on a sample of initiatives undertaken priorly to the beginning of Paris 2024 Olympic Games. In Table 3, we resemble them accordingly to the structure adopted also for the two prior reports and with the same additional logic in terms of initiatives undertaken and engagement tools involved providing also a brief description of the aims.

Not directly related to the stakeholders engagement initiatives, we noticed the recognition of the award for an "exemplary" execution of the ISO 20400, after the voluntary French Standardization Association (AFNOR) audit process. This lead to the obtainment of the "AFAQ Focus CSR Responsible Purchasing" label with refers to the social and environmental-led attention to the non-financial performance in purchasing and procurement process. Within the attentioning of socio-environmental contribution to the transition in the organization of such Olympic Games, they already calculated that they halved the GHG emissions if compared to the average of London and Rio prior editions (Third sustainability report, 2024, p. 17 and p. 61). This evidence has been evaluated in the analysis of the overall context of sustainability, as well as the endorsement of socio-economic and environmental pillars related to the contribution of Paris 2024 Legacy and Sustainability Strategy to the Sustainable Development Goals (SDGs) (Third sustainability report, 2024, p. 56).

As part of the alignment with the sustainable oriented strategy, we report in Table 3 also the programs played worth refers to the fight against the marginalization of the diverse and integration of the people affected by disabilities. Those evidences fuel a debate still animated with refers to the accountability also those issues, strengthening the need to pursue this debate (Caccialanza & Barabaschi, 2024). In particular, also in the context of London 2012 Olympic Games, particular attention had been played in the defence of people with disabilities' (PWD) employability (e.g., Minnaert, 2014).

Complementary to the events and tools to ensure, or at least promote, inclusivity of all stakeholders, a mention on the accessibility of structures must be considered. In particular, Paris 2024 faced significant challenges in constructing 44 temporary and permanent competition infrastructures while meeting eco-design standards (third sustainability and legacy report, 2024, p. 98). The goal was to ensure 100% reuse, repurposing, or recycling of materials from temporary structures and recover 80-100% of waste from the Athletes' Village decommissioning. Accessibility was prioritized, following French regulations and International Paralympic Committee guidelines. Measures also focused on minimizing environmental impacts, using existing venues, applying circular economy principles, protecting ecosystems, and enhancing resilience to climate risks through temporary cooling solutions and other protective features (He et al. 2022). In fact, the

Table 2. The level of stakeholder engagement - Interim phase (November 2023).

Dimensions of stakeholder engagement	Engagement tools or initiatives and involved stakeholders (IS)	Aim and description of the engagement tool
Stakeholder recognition	Qualitative survey. IS: Institutional organizations	Understanding of the environmental contribution potentially played by the organization
	Quantitative survey. IS: Staff	Understanding of the environmental contribution potentially played by the organization
Stakeholder support	"I'mPOSSIBLE", IS: children	It is an educational tool produced by the International Paralympic Committee for teachers in the form of a lesson plan
	The film "The Colour of Victory", IS: 350,000 schoolchildren	Those children have been able to access the film as a result of Paris 2024 making it available to watch free of charge
	The "Bouger Plus" (Move More!) strategy, IS: Stakeholders in general	This project increase the awareness of importance of physical activity and sport in their daily lives
Stakeholder dialogue	Generation 2024 Bulletin, IS: Educational community, the sporting movement, pupils and students	Registered members of the platform receive a quarterly newsletter and the stakeholders are all showcased in a section listing the projects they run on a daily basis
	The "Generation 2024" community. IS: Volunteers	Volunteering and civic service missions are promoted in close collaboration with the French National Olympic Sports Committee (CNOSF) and other federations

Sources: Institutional website and social media channels, and second sustainability and legacy report (2023).

design for Paris 2024 structures focused on sustainability by using lightweight, eco-friendly materials, ensuring thermal comfort, and prioritizing local suppliers to reduce environmental impact (third sustainability and legacy report 2024, p. 100). This effort is in line with part of the debate on the evaluation of the environmental impacts of

Table 3. The level of stakeholder engagement – Event-star-phase (May 2024).

Dimensions of stakeholder engagement	Engagement tools or initiatives and involved stakeholders (IS)	Aim and description of the engagement tool
Stakeholder recognition	the "Terrain d'Egalité" (Equal Playing Field) label; IS: Diverse and minorities	The development of a label for organizers of major international sporting events to encourage them to commit to equality and combat all forms of sexist and sexual discrimination and violence against the GBTQIA+ community
	Dedicated in-house team for accessibility of the Games; IS: People with disabilities (PWD)	
Stakeholder support	The "Integration through economic activity" program; IS: People with disabilities (PWD)	This activity aims to remove barriers to employment for the most disadvantaged by offering them work experience coupled with social and professional support. As a result, as of 2021, 170,000 people will have graduated from this program
	A "Territorialized Actions" Call for Projects for Professions of The Games; IS: Job seekers	Launched in 2022, this initiative offers additional training focused on local employment needs and those distant from the job market. So far, 21 projects under the theme "Professions of the Games" have created 1700 training opportunities
Stakeholder dialogue	"The Coach Climat"; IS: employees	This tool is available for all employees, and help them to estimate their annual carbon footprint, and choose actions to reduce their GHG emissions and see their effects
	The "Impact 2024 - From Stadium to Employment" events; IS: Job seekers	Since their launch, more than 36,000 job seekers and nearly 4500 companies have taken part in these job dating events across the country

Sources: Institutional website and social media channels, and third sustainability and legacy report (2024).

such mega-events and the necessity to engage this challenge while facing social inclusivity purposes (Mourgin et al., 2024).On the other hand, Paris 2024's approach to accessibility extends beyond physical spaces, yet its digital efforts since they are aligned with Accessibility Reference Framework for Administrations (RGAA) and Web Content Accessibility Guidelines. Those measures included accessible formats for the visually impaired and subtitles for content, but the need for continuous updates highlights a gap between ambition and full digital inclusivity for different audience and stakeholder's digital capabilities (third sustainability and legacy report, 2024, p. 109).

All things considered, the path for the comprehensive approach to sustainable management of Olympic events is an incremental approach adopted by prior editions and confirmed by this one. This incremental attitude and endorsement of sustainability issues is clearly stated in terms of stated preferences and priorities. For example, since the beginning of this reporting process, the first sustainability and legacy report (2021, p. 120) states that "*At the Games in London in 2012, women took part in all sports on the Olympic programme for the first time. In addition, the principle of gender equality was incorporated right from the recruitment of Organising Committee staff. These Games also showed that sport can change public attitudes towards disabilities. In Rio in 2016, clear measures were implemented to call out any instances of abuse or harassment. For the Tokyo Games, the event programme was modified to ensure equal representation of female and male participants. Paris 2024 plans to take this trend further and contribute in its turn by building even more inclusion, equality and solidarity into the model for the Games.*"

Albeit not all the despite efforts to present a comprehensive evaluation of the legacy and impacts of the Paris 2024 Games, there remains a risk of a partial perception, as stakeholder engagement largely excludes dissenting voices. The absence of opinions or protests against the Games, particularly from local communities exhibiting NIMBY (Not In My Backyard) attitudes, raises concerns about the 'balance' between institutional perspectives and those of other stakeholders. Reversely, this may be extended on the quality of the reporting according to the GRI 2016's principles, since this could result in an incomplete representation of the broader social impact, limiting the legitimacy and depth of the evaluation.

5 Conclusions and Research Agenda

This preliminary study aims to provide an overview of the Paris 2024 Olympic games propensity to disclose, while engaging the literature on the topic of sustainability accountability in the context of mega sports projects (Caccialanza et al., 2024). The contribution has traced the modalities (or attempt) of stakeholders engagement practices, analysing them with the perspective of the disclosure by Paris 2024 Olympic games organization committee on those managerial practices. For this purpose, a documental analysis has been conducted on the documents publicly available and the three sustainability reports published. Specifically, this interdisciplinary approach between accounting and project management literature supports the 'cross-fertilisation' perspective as proposed by the recent 'Manifesto for project management research' (Locatelli et al., 2023).

At the same time, this study presents some limitations. According to Bowen (2008, p. 32), the methodology used in this study may be affected by 'biased selectivity,' as documents and sustainability reports are often shaped by organizational policies. Future research can address this potential limitation in two main ways. First, as noted in the methodological section, extending the scope to include a wider range of mega-sport events or Olympic Games could provide insights into recurring limitations in how various material topics are discussed and disclosed. This broader analysis may also reveal the influence of others factors, such as the cultural one (Wang et al., 2023), given the international context of these events. Second, a more refined approach to mitigating selectivity would be to increase the diversity of documents – and data sources overall -analysed or to incorporate interviews with both internal and external stakeholders.

In line with this perspective and recognizing the limitations of this chapter, we suggest several directions for future research. Building on the findings, further studies should go beyond descriptive assessments and explore the more focused and granular aspects of event sustainability (Li et al. 2024). In particular, it might be of interest to examine potential opposition groups or stakeholder categories that may raise concerns regarding the sustainability practices of large-scale events such as the Paris 2024 Olympic Games. In those documents we found no evidence of those movements and no disclosure about the way they managed dissent, albeit this is a recurrent problem when is involved the evaluation of the social impact of local communities (Bourbillères et al., 2023; De Nito et al., 2024) Investigating these concerns could provide an evaluation or discussion of the underestimated ethical dimension of such initiatives.

Another area for further inquiry involves considerations in data collection procedures, particularly when comparing official or institutional reports with the perspectives of local communities and other stakeholders. Future research could explore how these different perspectives form the understanding and measurement of sustainability efforts and endorsement procedures. Additionally, it would be interesting to compare the pre-event and interim-reports with the announced future editions on the evaluation of the (sustainability) impacts and Legacy report that are expected to be published in 2025 and 2029 (Interim Sustainability & Legacy Report, 2023, p. 34).

As final consideration, the limitations associated with relying solely on desk-based research need to be addressed as priorly anticipated. The absence of interviews with both internal and external stakeholders leaves important (external stakeholders') perspectives unexplored. Future research should incorporate more direct forms of engagement, such as interviews or questionnaires, to offer a more detailed understanding of the mechanisms and outcomes related to event sustainability. This would help to overcome the limitations inherent in relying exclusively on secondary sources.

References

1. Ahmed S, Darwish AS, Farrell P (2022) Benchmarking sustainability of megaprojects: a review. In: Sustainable Energy Development and Innovation: Selected Papers from the World Renewable Energy Congress (WREC) 2020. Springer International Publishing, Cham, pp 743–753
2. Aksoy L, Banda S, Harmeling C, Keiningham TL, Pansari A (2022) Marketing's role in multi-stakeholder engagement. Int J Res Mark 39(2):445–461
3. Al Hallaq A, Ninov I, Dutt CS (2021) The perceptions of host-city residents of the impact of mega-events and their support: The EXPO 2020 in Dubai. J Policy Res Tour Leis Events 13(3):374–396

4. Babaei A, Locatelli G, Sainati T (2023) Local community engagement as a practice: an investigation of local community engagement issues and their impact on transport megaprojects' social value. Int J Manag Proj Bus 16(3):448–474
5. Barratt M, Choi TY, Li M (2011) Qualitative case studies in operations management: trends, research outcomes, and future research implications. J Oper Manag 29(4):329–342. https://doi.org/10.1016/j.jom.2010.06.002
6. Bourbillères H, Gasparini W, Koebel M (2023) Local protests against the 2024 Olympic Games in European cities: the cases of the Rome, Hamburg, Budapest and Paris 2024 bids. Sport Soc 26(1):1–26
7. Bourdin A, Dang V, Idt J, Meunier JM (2024) The Paris 2024 Olympic Games, steering a "Major Project" like any other? In: Planning the Paris 2024 Olympic and Paralympic Games. Singapore, Springer Nature Singapore, pp 55–68
8. Bowen, G. A. (2009). Document analysis as a qualitative research method. Qual Res J 9(2): 27–40. https://doi.org/10.3316/QRJ0902027
9. Brunet M (2021) Making sense of a governance framework for megaprojects: The challenge of finding equilibrium. Int J Proj Manag 39(4):406–416
10. Brookes NJ, Locatelli G (2015) Power plants as megaprojects: using empirics to shape policy, planning, and construction management. Util Policy 36:57–66
11. Bruzelius N, Flyvbjerg B, Rothengatter W (2002) Big decisions, big risks. improving accountability in mega projects. Transp Policy 9(2):143–154
12. Budzier A, Flyvbjerg B (2024) The Oxford Olympics Study 2024: are cost and cost overrun at the games coming down?. arXiv preprint arXiv:2406.01714
13. Caccialanza A, Cantoni F, Favari E, Mariani C (2022) Proposing a sustainability-centered approach to overcome critical issues in megaprojects. In: Sustainability and megaproject development. Routledge, London, pp 25–44
14. Caccialanza A, De Nito E, Canonico P, Favari E (2023) Mega-projects and social impact evaluation: the difficult (un) balanced inclusion of social needs in current practices. In: International Workshop "A Multidisciplinary Approach to Embrace Complexity and Sustainability in Megaprojects". Springer, Cham, pp 15–24
15. Caccialanza A, Barabaschi B (2024) Managing with the "Diverse": a gaming perspective for the promotion of a more inclusive and sustainability-oriented culture. In: Modarelli G, Rainero C, Amelio S (eds) Game-based education approaches to inclusive business management. IGI Global, Hershey, PA
16. Canonico P, De Nito E, Caccialanza A (2024) Sustainability disclosure and legacy in sports mega-events. In: eBook Proceedings IFKAD 2024, (Madrid, Spain, 12-14 June 2024), ebook IFKAD, Madrid, Spain
17. Cantarelli CC, Genovese A (2021) Innovation potential of megaprojects: a systematic literature review. Prod Plan Control:1–21
18. Chappelet J, Junod T (2006) A tale of 3 Olympic cities: what can Turin learn from the Olympic legacy of other alpine cities? In: Torres D (ed) Major sport events as opportunity for development. Valencia Summit proceedings
19. Corazza L, Torchia D, Cottafava D, Tipaldo G (2022) Considering the social and economic implications of infrastructure megaprojects: theoretical contributions, practical challenges and managerial implications. In: The impact of corporate social responsibility. corporate activities, the environment and society. Routledge, London, pp 1–30
20. Crawford P, Bryce P (2003) Project monitoring and evaluation: a method for enhancing the efficiency and effectiveness of aid project implementation. Int J Proj Manag 21(5):363–373
21. Del Cerro Santamaria G (2019) Megaprojects, development and competitiveness: building the infrastructure for globalization and neoliberalism. Athens J Soc Sci 6(4):263–290

22. De Nito E, Caccialanza A, Canonico P, Favari E (2024) Analyzing the role of social value in megaprojects: toward a new performance framework. Meas Bus Excell 28(2):243–254. https://doi.org/10.1108/MBE-09-2023-0138
23. DiVito L, Ingen-Housz Z (2021) From individual sustainability orientations to collective sustainability innovation and sustainable entrepreneurial ecosystems. Small Bus Econ 56(3):1057–1072
24. Dyer R (2017) Cultural sense-making integration into risk mitigation strategies towards megaproject success. Int J Proj Manag 35(7):1338–1349
25. Eisenhardt KM (1989) Building theories from case study research. Acad Manage Rev 14(4):532–550
26. Flyvbjerg B (2014) What you should know about megaprojects and why: an overview. Proj Manag J 45(2):6–19
27. Flyvbjerg B, Stewart A, Budzier A (2016) The Oxford Olympics Study 2016: cost and cost overrun at the games. arXiv preprint arXiv:1607.04484
28. Flyvbjerg B, Budzier A, Lunn D (2021) Regression to the tail: why the Olympics blow up. Environ Plann A Econ Space 53(2):233–260
29. Flyvbjerg B (2017) Introduction: The Iron Law of Megaproject Management. In: The Oxford Handbook of Megaproject Management, Chapter 1. Oxford University Press, Oxford, pp 1–18
30. He Q, Tian Z, Wang T (2022) Performance measurement methods in megaprojects: An analytical review. Int J Proj Manag 40(6):634–645
31. Ketokivi M, Choi T (2014) Renaissance of case research as a scientific method. J Oper Manag 32(5):232–240. https://doi.org/10.1016/j.jom.2014.03.004
32. Landis JD (2022) The megaproject challenge. In: Megaprojects for megacities: a comparative casebook, p 1
33. Leong C (1991) Accountability and project management: a convergence of objectives. Int J Proj Manag 9(4):240–249
34. Li Y, Wang M, Locatelli G, Zhang Y (2024) Navigating the future of megaprojects sustainability: a comprehensive framework and research agendas. Int J Manag Proj Bus 17(3):533–561
35. Locatelli G, Ika L, Drouin N, Müller R, Huemann M, Söderlund J, Gerardi J, Clegg S (2023) A Manifesto for project management research. Eur Manag Rev 20(1):3–17. https://doi.org/10.1111/emre.12568
36. Lovallo D, Cristofaro M, Flyvbjerg B (2023) Governing large projects: a three-stage process to get it right. Acad Manag Perspect 37(2):138–156
37. Mougin L, Bougault V, Racinais S, Mountjoy ML, Stephenson B, Carter S et al (2024) Environmental challenges facing athletes, stakeholders and spectators at Paris 2024 Olympic and Paralympic Games: an evidence-based review of mitigation strategies and recommendations. Br J Sports Med 58(15):870–881
38. Modell S (2005) Triangulation between case study and survey methods in management accounting research: an assessment of validity implications. Manag Account Res 16(2):231–254
39. Moggi S, Giacomo F, Elisa I (2024) Enhancing value creation through stakeholder engagement in sustainable mega-events: the case of Expo 2020 Dubai legacy. Bus Ethics Environ Responsibil
40. Millar J, Mueller F, Carter C (2024) Grassroots accountability: the practical and symbolic aspects of performance. Account Audit Account J 37(2):586–607
41. Minnaert L (2014) Making the Olympics work: interpreting diversity and inclusivity in employment and skills development pre-London 2012. Contemp Soc Sci 9(2):196–209

42. Müller M, Wolfe SD, Gogishvili D, Gaffney C, Hug M, Leick A (2022) The mega-events database: systematising the evidence on mega-event outcomes. Leis Stud 41(3):437–445. https://doi.org/10.1080/02614367.2021.1998835
43. Paris 2024 Organising Committee for the Olympic and Paralympic Games. (2021, 30 agosto). The legacy and sustainability plan for the Paris 2024 Olympic and Paralympic Games [PDF]. https://olympics.com/en/paris-2024/legacy-plan-en
44. Pitsis A, Clegg S, Freeder D, Sankaran S, Burdon S (2018) Megaprojects redefined–complexity vs cost and social imperatives. Int J Manag Proj Bus 11(1):7–34
45. Rook W, Heerdt D (eds) (2023) The Routledge handbook of mega-sporting events and human rights. Taylor & Francis, London
46. Salama M (2021) Event Project Management: Principles. Goodfellow Publishers Ltd
47. Stake RE (2005) Qualitative case studies. In: Denzin NK, Lincoln YS (eds) The SAGE handbook of qualitative research, 3rd edn. SAGE, pp 443–466
48. Sydow J, Braun T (2018) Projects as temporary organizations: an agenda for further theorizing the interorganizational dimension. Int J Proj Manag 36(1):4–11
49. Söderlund J, Sankaran S, Biesenthal C (2017) The past and present of megaprojects. Proj Manag J 48(6):5–16
50. Fondazione Milano Cortina 2026 (2023) Sustainability, impact and legacy Report - 2023: The strategy and planning phases. https://milanocortina2026.olympics.com/s3fs-public/documents/2025-01/Sustainability%2C%20Impact%20and%20Legacy%20Report%202023.pdf
51. Too EG, Weaver P (2014) The management of project management: a conceptual framework for project governance. Int J Proj Manag 32(8):1382–1394
52. Varrel A, Kennedy L (2011) Mega-events and megaprojects. Policy Brief 3:1–4
53. Vickerman R (2017) Wider impacts of megaprojects. In: The Oxford handbook of megaproject management, p 389
54. Wang G, Wu P, Wu X, Zhang H, Guo Q, Cai Y (2020) Mapping global research on sustainability of megaproject management: a scientometric review. J Clean Prod 259:120831
55. Wang Z, Zhang Y, Zheng K, Zeng R, Yuan H, Liu J (2023) A review of mega-project management research from an organization science perspective: current status and future directions. In: Developments in the built environment, p 100254
56. Xue J, Shen GQ, Yang RJ, Wu H, Li X, Lin X, Xue F (2020) Mapping the knowledge domain of stakeholder perspective studies in construction projects: a bibliometric approach. Int J Proj Manag 38(6):313–326
57. Yin RK (2018) Case study research: design and methods, 6th edn. SAGE Publications, Thousand Oaks, CA. www.books.google.co.uk/books?id=OgyqBAAAQBAJ

Infrastructure Refurbishment: The Tunnel Renewal Strategy from the Knowledge to Renovation of the Asset

Daniele Di Fiore(✉), Carlo Alessio, Lapo Baccolini, Fabrizio Colla, and Marilisa Conte

Tecne Gruppo Autostrade per l'Italia S.p.A, Rome 00159, Italy
daniele.difiore@tecneautostrade.it

Abstract. In 2020, the Italian Ministry of Infrastructure promoted a massive assessment on structural tunnel, followed in 2022 by the issuance of a new, more structured and risk-based assessment standard which has enabled to plan a new concept to refurbishing works of the tunnels along the network. Autostrade per l'Italia, responding to this high demand for maintenance work, mainly related to a significant part of the network built before the 1970s, has developed a new approach to renovate the existing tunnel asset, named Tunnel Renewal Strategy ("TRS"). TRS is a comprehensive approach which aims to achieve service life's renovation of existing tunnels through different re-lining solutions involving the demolition of a portion of the existing lining, the insertion of waterproofing and the construction of a new - thin high-performance structural shell. Over the last three years the design of the solutions has been developed and implemented through pilot projects. The construction industry was approached by engaging specialized players in the development of innovative solutions. Right now, about 40 tunnel projects have been designed for relining purpose and half of them are already completed (or on the way to be). In addition, over the last three years, many refurbishments and temporary measures have been carried out on the whole motorway network to improve safety and service levels for users.

Keywords: Tunnel renewal strategy · Tunnel asset management

1 Introduction

1.1 Overview

Italy has a complex transport system mainly due to its geography and orography. Moving people and goods in the most safe, fast and efficient way is nowadays crucial and strategic at the same time.

Road transport mode is historically the most widely adopted in the Country. In this picture the resilience of the motorway is one of the main objectives that Autostrade per l'Italia S.p.A. (ASPI), the main motorway Concessioner in Italy, is pursuing.

E. Favari et al. (Eds.): MeRIT 2024, LNCE 600, pp. 175–196, 2025.
https://doi.org/10.1007/978-3-031-92797-3_12

ASPI is actively involved in the maintenance and operation of the asset on behalf of the Ministry of Transport. The main purpose is to ensure travellers, as well as employees, safety and continuity of the operations within a competitive level of service.

An extraordinary assessment plan was launched in 2020 on the road tunnel side, triggering about 365 km of ASPI network (595 tunnels) (Fig. 1) [3].

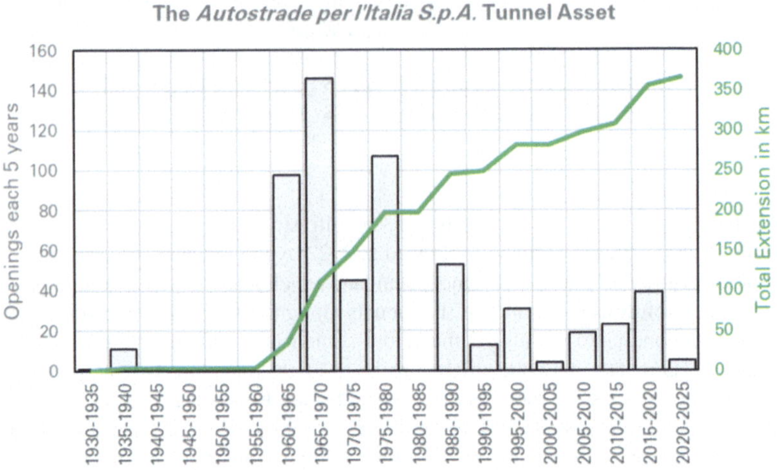

Fig. 1. The development of ASPI tunnel asset each 5 years.

The 595 existing motorway tunnels managed by Autostrade per l'Italia have several diameter and number of lanes (from 1 to 4). Most typical are 2 lanes tubes (5.05 m in radius) and 3 lanes ones (6.63 m in radius). They vary in length from 50 m to 8.7 km (average 600 m) for a total of 365 km.

In 2020, the assessment program was mainly characterized by a reactive approach to maintenance (phase 1) due to few adverse events, one of which is unfortunately well known.

Tunnels were visually inspected and tested through laser scanning (LSC) (Fig. 2) and ground-penetrating radargrams (GPR) as well as cored to compressive strength each 50 m.

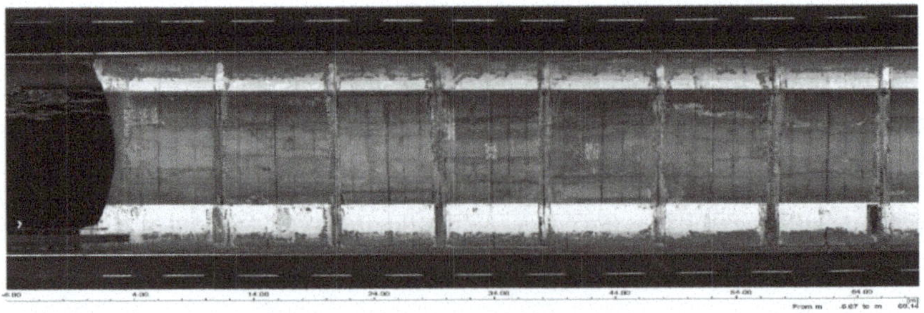

Fig. 2. Example of laser scanning result.

Many tunnels were visually inspected simultaneously along the motorway. At that time, due to the pandemic, the level of traffic was the lowest ever recorded since many decades and this facilitated construction sites, minimizing the effect on users of the infrastructure (Fig. 3).

Fig. 3. Detailed visual inspection, hammering and laser scanning.

The exceptional tunnel assessment program is based on a series of activities, mainly related to the inspections, maintenance, design and construction, totaling approximately € 995 million. Civil works, related to refurbishment and modernization of the tunnel linings, are 80% of overall cost.

The initial reactive maintenance was leading to treatment of structural defects in a short time frame (3 to 6 months) opening several construction sites at the same time, increasing the user related traffic risk when approaching the tunnel, something no more sustainable to date.

The structural renewal plan is a key part of the Tunnels Assessment's overall program to achieve the full knowledge of asset conditions and renew the tunnels within the existing ASPI motorway network. The Tunnels Assessment's overall program comprises systematic surveillance and detailed inspections, ordinary maintenance works, structural and geomechanical surveys, risk analyses and safety evaluations of tunnels vulnerability. Strictly based on the outputs of this assessment process, the main intervention strategy corresponding to tunnels relining, in accordance with Tunnel Renewal Strategy (TRS), is combined with Extraordinary Maintenance Interventions (IMS) and Structural Reinforcement Interventions (IRS). These approaches aim to ensure a proper preservation of the structures, gradually renewing their useful life, increasing their resilience by improving safety and structural performance. These efforts are in line with the Group's broader vision of a resilient and safe infrastructure.

1.2 The National Code for Existing Tunnels

A transition from reactive to preventive maintenance approach (phase 2) was then needed among the Organizations and partially promoted by the new Guidelines:

- 23rd August 2022 "Guidelines for existing tunnel's risk and safety evaluation and monitoring system installation", referred to as "LLGG" in the following sections [1].

The new code comes spread in layers (0 to 4) characterized by an increasing level of detail:

- **Level 0** – Asset information such as location, length, lining material, construction age etc.;
- **Level 1** – Asset conditions in terms of scored defect detected during the visual inspection. In this phase, extension (k_1), magnitude (k_2) and severity (G) of the defects are collected in forms for further analyses;
- **Level 2** – The levels 0 and 1 gathered information feeds a process whose purpose is to assign a simplified risk score to each 20 m long segment of the tunnel for each thematic area (structural global and local, geological, seismic, hydraulic and transport related). The score can be low, mid-low, mid-high or high (Fig. 4).

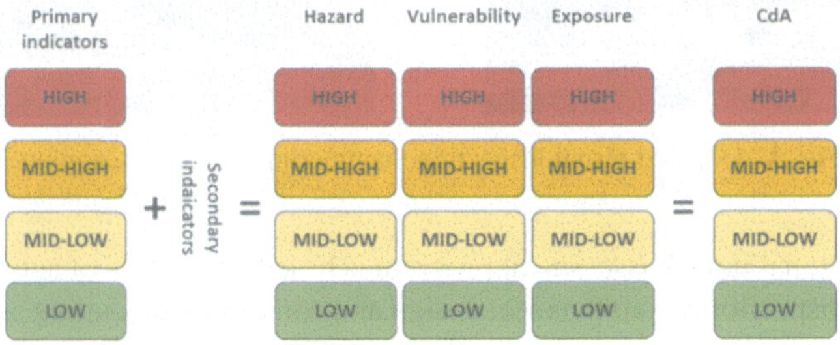

Fig. 4. Definition of the simplified risk score "Classe di attenzione".

The six thematic areas (except for hydraulic) are then combined into a global risk rate. Level 2 defines the visual inspection frequency of the asset, by the risk rate associated to structures (both for global and local mechanisms).

- **Level 3 & 4** – Whenever risk rate coming from level 2 is mid-high a preliminary safety evaluation is performed to better investigate the parameters which lead to such rate of risk in the specific thematic area. If the risk rate is confirmed engineering judgement and calculation are put in place to identify the mitigation measures, or eventually monitoring systems, needed to keep the risk acceptable (Level 4). When the risk rate is higher Level 3 is glossed over. At Level 4, the engineered process to residual level of safety evaluation includes calculation and approaches, in some cases, directly derived from the past.

The LLGG multilevel approach helps Organisations like ASPI, with a very large asset portfolio, in prioritizing maintenance and expenditures.

In the following sections the assessment program is described starting the treatment strategies with an insight to the asset management and a focus on the tunnel renewal strategy with some return of experience.

1.3 The Intervention Strategies

Level 4 purpose is to highlight the criticalities and possible solutions in terms of monitoring systems or intervention strategies. Tecne S.p.A., engineering company of Autostrade per l'Italia S.p.A., tends to group intervention strategies with the effect these have on asset conditions:

- IMS – is the treatment approach closer to extraordinary maintenance "*Intervento di Manutenzione Straordinaria*". The aim is to reduce vulnerability for mainly local mechanisms of failure without increasing the strength of the structure (Fig. 5).
- IRS – stands for structural reinforcement "*Intervento di Rinforzo Strutturale*". In this, the aim is to reduce vulnerability for global mechanisms and increase the structural strength of the system (Fig. 6).
- TRS – stands for "*Tunnel Renewal Strategy*". The renewal solution is aimed to extend the tunnels life cycle of at least additional 50 years, through the construction of new lining, able to replace the structural function of the existing one and to ensure waterproof condition and suitable performance under seismic events (Figs. 7, 8 and 9).

Fig. 5. IMS treatment solution for cracks.

Fig. 6. IRS treatment solution to reinforce the tunnel crown by DCP rock bolts and invert by reinforced cast in place concrete.

Fig. 7. On the left side Colle right tunnel (A01) – relined by cast in place concrete; on the right side Manfreida left tunnel (A26) – relined by high performance sprayed concrete.

Following Sect. 3 focuses on the whole TRS approach, from its objectives ad requirements up to the description of typological interventions and an extensive case history, passing through the implemented principles regarding the involvement of construction market, sustainability and traffic management .

The optimal solution to be implemented mainly depends on Organization culture about preventive maintenance, existing lining conditions, impact on traffic, time, costs, waterproofing effectiveness, job-site organization, reusability of the solution, propensity to innovation.

Fig. 8. San Bernardo right tunnel (A12) – relined by steel liner plates.

Fig. 9. Poderuzzo right tunnel (A01) – relined by concrete precast segments.

On a higher organizational level, decision making on tunnel maintenance strategy is moving to a more robust asset management approach also by developing a dedicated tool to compare several alternative intervention scenarios in terms of risk, costs and residual value of the asset.

2 Asset Management for Tunnels

The Organization is focused on the use and further development of a decision supporting tool for tunnel asset management [6] (Fig. 10).

The methods and approaches herein summarized are applicable to any kind of civil asset then, extension to bridges and viaducts are under investigation as well.

About tunnels, the purpose is to base the tool on the new Guidelines, which represent the common rules for all the Owners/Concessioners of road tunnel asset in Italy. The new regulation is promoting further actions to ensure the implementation of a more reliable approach to asset management "*Si raccomanda inoltre che detti modelli possano essere anche aggiornati in tempo reale, costituendo quindi un banca dati aggiornata per le necessarie azioni di asset management*".

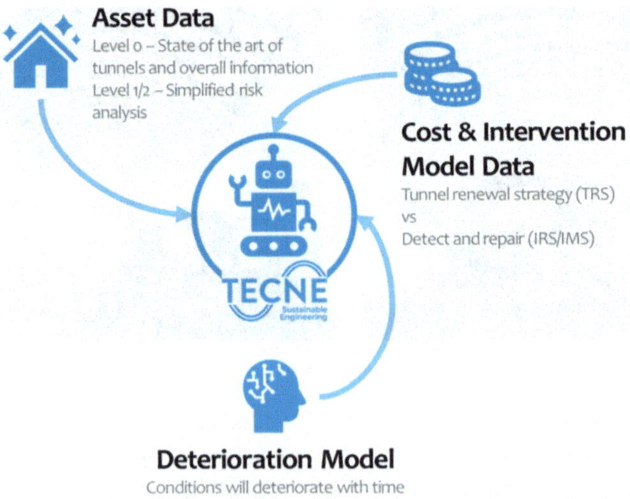

Fig. 10. Tool input

The decision support tool is feed by the information required by the LLGG among the ones used for the multilevel analyses. Data coming from levels 0, 1 and 2 are necessary to assess the condition of the asset at time 0.

Deterioration models are then implemented to evaluate the foreseen condition of the asset in the most reliable way, compatible with knowledges and data available at the time of the analysis. The aim is to have a set of models dedicated to specific assets (ie.customized by construction age, liner materials or waterproofing technology). Cost and intervention models are included to compare alternative scenarios where Tunnel Renewal Strategy (TRS) or Detect and Repair approaches (IRS/IMS) are possible. The decision support tool is aimed to compare several parameters such as current and foreseen monetary value of the asset, Capital (CapEx) and Operative (OpEx) discounted expenditures (Fig. 11), disruption, environmental impact, effects on reputation etc., in different operating scenarios (including no intervention at all).

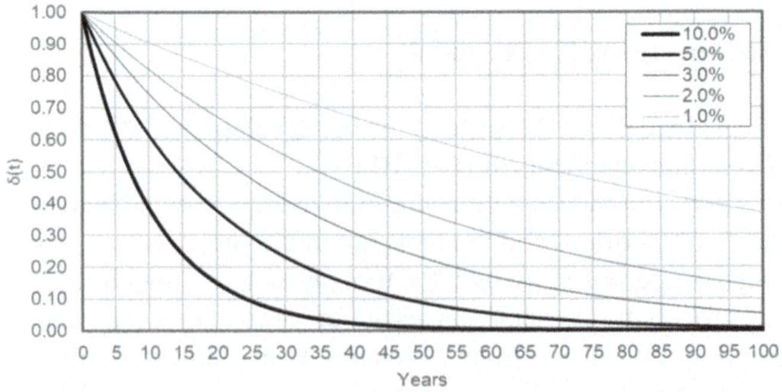

Fig. 11. Discount factor for different discount rates vs time.

Inputs feeds Outputs and vice versa into a never-ending loop of asset lifecycle.

Developing a Life Cycle Cost Analysis (LCA) tool brings multiple advantages to the Organization. Herein some of them:

- No bias in the decision-making processes concerning intervention planning and design;
- Easy access to the assumptions underlying the algorithm;
- Data driven processes based on the state of the art of the asset management industry.

At this stage, the tool is a web app version developed in strict collaboration with the Polytechnic University of Tourin. It works on four objectives functions that are optimized to have the optimum or sub-optimum treatment solution as a combination of IMS, IRS and TRS.

3 The Tunnel Renewal Strategy

In less than two years from the beginning of the second phase (2022), many relining typological sections have been designed, tested and the application of the related pilot projects has been completed. The design and testing stages have been extended to include members of the construction company and all technicians involved in the process 3.

Since 2023, projects have been developed in recurrent applications, taking into account the lessons learned gained during the pilot stage (Fig. 12).

Fig. 12. San Fermo right tunnel (A09) – field test of shotcrete application.

The Tunnel Renewal Strategy is still ongoing, but here the return of experience from the major civil engineering works done until March 2024 are presented (Fig. 13).

3.1 Objectives and Requirements

Due to the early-stage tunnel assessment experience mainly based on reactive maintenance, the TRS approach was developed since end of 2021 by the following main steps:

n.	Tunnel	//	L (m)	Typological	Status
1	Castello I left	A10	52	Steel liner plates	Complete
2	S. Fermo rigth	A09	701	Shotcrete	Complete
3	Colle Marino left	A14	952	Precast liner	Complete
4	Olimpia right	A26	903	Conc cast in place	Complete
	Manfreida left	A26	787	Shotcrete	Complete
	Poggio right	A26	455	Shotcrete	Ongoing
5	Lagoscuro right	A26	226	Shotcrete	Complete
	Ciutti right	A26	553	Shotcrete	Ongoing
	Setteventi right	A26	409	Shotcrete	Ongoing
	Colle right	A01	169	Conc cast in place	Complete
	Ragnaia II right	A01	128	Conc cast in place	Complete
6	Ragnaia I right	A01	63	Conc cast in place	Complete
	Croci di C. right	A01	840	Shotcrete	Ongoing
	Poderuzzo right	A01	216	Precast liner	Ongoing
	Del Fico right	A12	240	Precast liner	Ongoing
7	S. Bernardo right	A12	86	Steel liner plates	Complete
	S. Anna right	A12	860	Conc cast in place	Ongoing

Fig. 13. Status of the ongoing TRS projects.

1. analysis of the assessment program and technical solutions applied during 2020/2021 (phase 1);
2. collection of existing tunnel industry state of the art, including bibliographic research, looking for solutions adopted around the word for tunnel rehabilitation both in continuity of operation and closure;
3. definition objectives and requirements of relining works;
4. definition of a set of typological interventions and related design methods and verification criteria;
5. involvement of construction market to develop mechanized and technologically innovative solutions;
6. organizing field tests and application of pilot projects;
7. standardization and recurrent applications based on return of experiences coming from pilot projects;
8. continuous improvement.

The main objectives and requirements defined in the TRS context are:

- safety for operation, especially in case of alternation of works/traffic operation and works during traffic operation;
- safety for road-workers (protection, mechanization);
- minimization of impact on traffic (protection, mechanization);
- quickness of intervention (mechanization);
- high structural performance (pre-casting operations, high-performing concrete mix);
- installation of an effective and durable waterproofing system;
- sustainability and conservation of the asset by recycling, use of green machine and materials

Key role to ensure safety conditions and reduce time and traffic impact is played by the minimization of the existing concrete milling, especially when a not negligible stress state is detected in the lining during detailed inspection. This allows also to reduce the supports and soil treatment needed to fulfill the required safety standard during concrete demolition, maintaining low uncertainties regarding rock mass stability (Fig. 14).

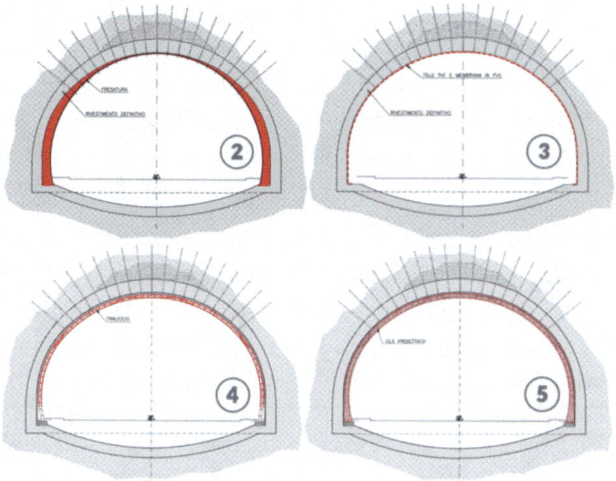

Fig. 14. TRS main phases – (2) milling, (3) waterproofing, (4) reinforcement installation, (5) final lining.

The minimization of existing concrete milling can be obtained by lowering the road pavement and reducing new lining thickness resorting to high performance materials.

Once the new lining is built, the maintenance of its structural capability is ensured by the mandatory installation of a waterproofing system. Drainage system is mandatory to take out water from deterioration root causes during the operational stage.

Every project is accompanied by environmental performance of the works, assessed by LCA methodology, which has become an additional item for designing and choosing between engineering solutions.

3.2 Involvement of Construction Market

As part of TRS, the construction market was approached by engaging specialized players by issuing the challenge regarding the development of innovative solutions.

The search for these solutions was mainly directed in the area of materials, seeking innovative materials with high structural performance, as well as in the area of mechanization for the installation of the components of the new lining shell. Not Disclosure Agreements have been signed with the selected companies and suppliers to ensure the exclusive use of the information shared in the design phase, then followed by specific agreements in order to manage future patent developments of original solutions as well.

3.3 Typological Interventions

Nine different typological interventions were defined (Fig. 15).

:

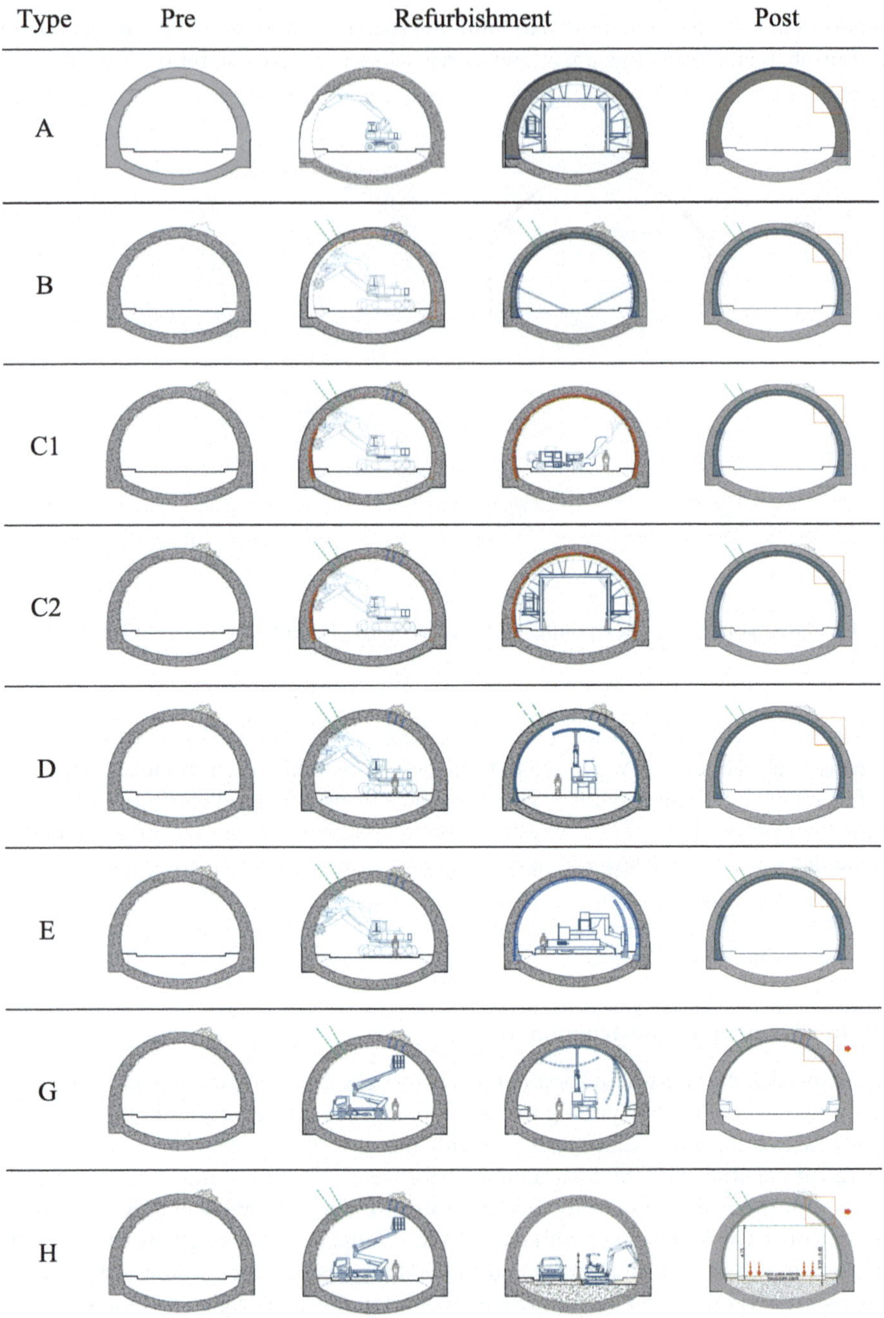

Fig. 15. Typological intervention conceptual schemes.

A. complete demolishing of existing lining and cast in place of new concrete shell;

B. partial existing lining milling and concrete relining casted in place by precast concrete shell formworks;

C1. partial existing lining milling and relining by high performance shotcrete;
C2. partial existing lining milling and concrete relining by casted in place by modular steel formwork;

D. shallow existing lining milling and relining by implementation of liner plates;
E. existing lining milling and relining by implementation of precast concrete segments;
F. steel liner coupled to existing lining;
G. installation of steel ribs and shotcrete, with or without existing lining milling;
H. platform remodulation (lowering) to avoid existing lining milling combined with inner lining based on previous typological interventions.

Each intervention has been detailed evaluated to define its proper field of application based on a pros and cons analysis and the specific design methods and verification criteria were developed correspondingly

3.4 Sustainability

The focus on the existing tunnels and the intervention strategies adopted cover several aspects of sustainability, including the protection of natural resources and biodiversity, ensuring efficient use of resources, safety during the construction phase and maintenance activities, while reducing the duration of works, minimizing the impact on traffic and extending the residual useful life of tunnels and ensuring high structural performance and low maintenance costs in the future [4].

Any treatment strategy applied to tunnel's segments and/or associated parts, avoid negative impacts on environmentally and scenically valuable areas outside the tunnel. This also ensures that natural lands are not converted for infrastructural use. No new abiotic sector is exposed to risk or stress, protecting organisms and their habitats and maintaining ecosystem functionality. In addition, the decision to work exclusively within the motorway boundary completely avoids the use of virgin soil, including temporary occupation during refurbishment and renewal activities.

From an environmental management perspective, the construction site do not introduce xenobiotic substances into the natural environment. Within the job site boundaries, the only certain chemical compounds used are additives for construction materials. It should be noted that throughout the entire duration of the work, all necessary precautions and good practices are adopted to avoid accidental discharge of potentially polluting liquids and to ensure protection from pollution potentially generated directly or not directly by construction activities.

As part of the Tunnel Renewal Strategy (TRS), a technology working group has been established involving various companies and suppliers (Sect. 3.2) to facilitate the implementation of initiatives aimed at developing innovative technologies. These technologies are intended to optimize construction processes, speed up implementation and use high performance construction materials that offer durability and strength. Innovative construction techniques enable the use of sustainable building materials, resulting in lower atmospheric carbon dioxide emissions compared to traditional ones. For example, high performance shotcrete and pre-assembled reinforcement cages reduce intervention times. Revitalizing existing tunnels rather than creating new ones has a positive social impact, avoiding the need for expropriation of new areas and potential impacts on the community. The technical solutions adopted aim to speed up and industrialize the work, reducing the impact of construction sites on motorways while maintaining acceptable levels of vehicular traffic flow and time savings. The plan also includes a water-saving strategy, by modernizing the collection and distribution system of a natural water source around the tunnels (where available), which supplies water to nearby homes.

Finally, the ARGO Gallerie platform is used to manage the surveillance activities of all the 595 tunnels in the network. It's a Tunnel Management System designed to streamline and optimize maintenance processes by collecting data from each tunnel into a digital inventory. This system allows for the creation of detailed inspection plans and intervention scheduling in a reliable, paperless digital environment.

3.5 Innovative Traffic Management Systems

Road Zipper technology has been used to make busy motorway sections more efficient, safe and functional. Tested for the first time in Liguria, within the Manfreida tunnel renovation project, this technology helps to manage the impact of a construction site on users (Fig. 16).

This machine it is around 3 m wide and more than 30 m long, as the name suggests, by physically moving a specially designed concrete barrier, can radically modify the layout of the motorway in seconds.

Based on the changing needs of peak traffic, the machine can zip the opposite carriageway up and down multiple times a day, reconfiguring the road without having to build a new lane. For adding new lanes, in fact, you have years of construction, sometimes you need to get environmental approvals, sometimes you need to buy lands to expand the motorway and in some cases the space is not enough.

Road zipper is a big innovation since it can mitigate congestion due to a tunnel closure with the existing opposite direction motorway capacity.

To change the layout, the road zipper uses a system of rubberized convey wheels. As the operator drives forward, it gently lifts and shifts the barrier diagonally across the underside of the machine.

Crucial to the system is the barrier itself, made up by hundreds of interlocking one meter section. The seven hundred kilos t-shape elements are linked and articulated by metal pins making the barrier an ultra-strong but flexible chain.

By travelling at a speed of up to eight kilometres per hour, the machine can zip a two kilometres barrier in just 15 min.

Fig. 16. Road zipper working at Manfreida left tunnel (A26).

As well as keeping traffic moving, it creates a safe divide between vehicles which are travelling towards each other at danger speed, something that humble traffic cones cannot do. Finally, road zipper enhances timesaving, cutting months on the schedule and allowing re-opening in advance.

"Protective shields" are also often used to separate the site from the motorway. This solution allows the tunnels to remain operational (reducing the number of lanes) while work is carried out behind the shield, ensuring safety and a sufficient level of service for users. Similarly, the use of "heavy" steel formwork in the relining operations allows to pour concrete during the night shift and left it in place during the day to ensure the required level of safety for users.

Sometimes the success of a project lies in the combination of several rehabilitation activities on the same motorway section. To achieve this, work in the tunnels is sometimes postponed, with temporary measures and monitoring systems installed to keep the area safe while the works on other assets are optimized.

3.6 The First Case-History of Full Lining Reconstruction

The Colle Marino left project [2] [5], along A14 Bologna-Taranto motorway, is the first case, in Italy, of complete reconstruction of existing tunnel lining, from invert to crown, over the entire length (960 m) (Figs. 17, 18 and 19)).

The technical complexity due to the invert demolition in clayey geological context, with overburden of up to 70 m and adjacent tunnel open to traffic, required firstly an extensive geotechnical-survey campaign and systemic concrete-structures investigations, from which significant linings working rates emerged, especially at maximum overburden in higher plasticity clays.

Fig. 17. Colle Marino left tunnel (A14) – existing and new linings.

Fig. 18. Colle Marino left tunnel (A14) – invert reconstruction.

The project therefore provided for a careful planning of invert demolition and reconstruction operations, arranged by segments of various lengths at proper spacing, with the adoption of specific demolition-technologies for early-assessment of invert actual working-conditions and structural-disturbance minimization (Fig. 18).

Specific monitoring procedures and emergency measures have been also established to counteract deformations and existing lining damaging. After complete invert-reconstruction using PP macro-fibers reinforced concrete, temporary re-opening to traffic preceded crown-piers reconstruction adopting pre-cast arch slabs with SCC back-pouring fully waterproofed by TNT + PVC membrane (Fig. 19).

Construction works, carried out during several H24 tunnel-closure periods (almost 7 months of net activity), included considerable road platform lowering (65 cm) thanks

Fig. 19. Colle Marino left tunnel (A14) – final look.

to the absence of altimetric constraints at portals. This solution reduced milling-depth at crown thus strongly mitigating risks associated to demolition of existing concrete lining extensively guarded by temporary steel nailing.

General speeding-up of production was therefore achieved. Industrialization of construction processes was ensured thanks to knowledge framework completeness and efficient site-organization. This valuable know-how allowed to plan similar works on adjacent tunnel by granting daily traffic opening, introducing significant technical optimizations for both invert and crown structures. The innovative TRWT system (Tunnel Renewal Work Train) will be adopted to meet workers-safety and automation of main construction processes (milling, waterproofing and high-performance concrete pouring), while ensuring appropriate level of service for motorway users and is more deeply investigated in section.

3.7 Steel-Liner Plates—Fast and Neat Solution in Short Tunnels

The Castello 1 left tunnel is located along the A10 motorway [2]. Its construction dates back to the second half of the 1950s, i.e. the time when the single-carriageway motorway section between Genoa Prà and Albisola was built. The intrados profile consists of a monocentric crown with a radius of 6.11 m, with sub-vertical about 2 m high.

Following structural investigations, it was clear that the original final linings were made of plain concrete, whose thickness goes from 20 cm and 120 cm. There is no invert.

In the year 2020, the tunnel was subject to deep inspection as part of the Tunnel Assessment Program put in place by ASPI. The defects identified through the above-mentioned inspection, which were further outlined through the structural investigations carried out on the final linings, required a safety temporary intervention consisting of an inner arch made on steel ribs and shotcrete realized on about 25 m (from Savona's side entrance), in addition to the installation of further supports (i.e., steel nets, metal sheets for water collection) in the remaining part of the tunnel.

The construction of the above-mentioned inner arch resulted in a significant narrowing of the roadway with a consequent reduction in the number of lanes from 3 to 2. Because of this, ASPI has considered the structural renovation of the final lining a priority, in order to make sure the whole roadway could be used again thanks to the demolition of the temporary arch and subsequent reconstruction of the final lining of the tunnel.

In summary, the intervention involved:

- the demolition of the inner arch;
- the partial demolition of the original final lining, over the entire profile development (including piers and walls);
- the equipping of the tunnel with new lining waterproofing system, drainage and collection of seepage water from the pile;
- the construction of a new final lining consisting of steel lining plates founded on newly reinforced concrete kickers;
- the full reconstruction of the pavement.

The new final lining retains the intrados profile of the original lining, thus reinstating it to the condition prior to the construction of the inner arch. At the newly built final lining is given a Nominal Design Life of 50 years, as conventionally defined in the Standards.

Basically, the technical solution consisting of liner plates was adopted because there was the need to create a new lining as thin as possible.

The presence of asbestos fibers in the original linings required the adoption of special technologies for the demolition of the lining, i.e. the use of the hydrodemolition technique with the reuse of process water (following filtering) in the first section of the tunnel, including the inner arch, and the milling operation implemented according to specific procedures to minimize dust dispersion in the remaining section of tunnel. The adoption of the hydro demolition technique in the first tunnel section was closely associated with the presence of the inner arch, which turned out to be intimately connected to the original lining by means of a system of metal hooks embedded in the concrete. This configuration clearly did not allowed to carry out milling operations, given the significant interference with the connecting hooks.

Liner plates consisted of 120 cm long standard modules made of three circumferential elements connected by means of 2 joint plates (i.e., bolted connection) equipped with a hinge to allow the transportation of the arch in a semi-closed configuration inside the tunnel, and the opening for installation.

After the installation, the individual modules were bolted along the core of the main lateral trusses, in parallel with the tunnel axis, ensuring longitudinal continuity to the system. Superfluid grout made with cement-based binder with high resistance to sulfates was used for back side injection (Fig. 20).

Fig. 20. Castello I right tunnel (A10) – relined by steel liner plates.

3.8 Relining by Sprayed Concrete—When Flexibility Is Needed

The San Fermo right tunnel is located in the northbound carriageway of the A09 motorway. The tunnel, which was built in the first half of the 1970s, is about 700 m long and has a two-lane carriageway with a total width of 9.5 m.

The structural renovation of the existing lining, for the entire length of the tunnel, was consisting of a high-performing structural shotcrete shell placed with the help of modular panels of electro welded mesh. The shell was set on newly built reinforced concrete kickers. The new final lining was preceded by the installation of a waterproofing system consisting of a TPO sheet protected by a double layer of TNT (extrados/intrados). The tunnel was equipped with a network to collect infiltration water from the rock mass. The work included the reconstruction of the platform water collection system, fire-extinguishing system, SOS, as well as the construction of a new lighting system.

The reconstruction of the final lining did not change the intrados profile of the tunnel, so the functional section and the minimum clearance on lateral lanes remained unchanged. At the newly built final lining is given a Nominal Design Life of 50 years.

As a matter of fact, the tunnel's original unreinforced lining was subject to an extremely modest state of stress, which was essentially attributable to its self-weight.

Therefore, the conditions allowed for the adoption of a lining which does not require the use of reinforcement in order to meet specific structural requirements both under static load and in the presence of seismic events (also given site low-seismicity).The adoption of the technological solution based on the use of shotcrete instead of cast-in-place concrete for the construction of the final lining is due to the greater versatility of the construction process, which allows to manage in a rapid and effective manner short-term construction sites and/or constructions sites where works are carried out on a non-continuous basis, as is the case of the San Fermo Tunnel.

The work plan was divided into three macro-phases. In Phase 1, since it was necessary to use a system that can be rapidly made available on the market, Ram – Arch® system was used. This system, which was developed in Great Britain, has been widely applied in interventions of renovation of the final linings of existing tunnels. It consists of modular elements of electro-welded mesh placed according to three-dimensional geometry, specifically equipped with V-shaped corrugations towards the tunnel intrados which act as ribs for the creation of a self-supporting mesh being stiff enough to ensure the stability of the system during the shotcrete application phase. For Phase 2 and 3, modular fiberglass mesh cages were used, since there was time to design systems of equivalent fast installation (Figs. 21 and 22).

Fig. 21. San Fermo right tunnel (A09) – waterproofing system installation.

The project involved the consolidation and the systematic plugging of all the cavities being in the linings' outline by means of cementitious mortars for injections characterized by the rapid development of mechanical strengths. Following the completion of the filling of the back voids and the consolidation phases of the original lining, it was necessary to proceed with the radial consolidation work of the rock mass around the concrete lining of the tunnel.

The structural renovation of the final linings involved the construction of a structural shotcrete shell, 30 cm thick, after the milling of the existing claddings and the installation of the waterproofing system.

Fig. 22. San Fermo right tunnel (A09) – relining by high performance shotcrete.

4 Conclusions

Tunnel assessment and renewal is an ongoing process for Autostrade per l'Italia S.p.A. The amount of information and experience acquired by its Engineering company (Tecne S.p.A.) and all the Operators involved in the activities, is constantly growing. The monetary and economic value of the asset make any effort needed and worthy.

The reactive approach to maintenance has been overtaken and the Tunnel Renewal Strategy is now starting the standardization and recurrent applications stage to extend the tunnels life cycle of at least additional 50 years.

Beyond that, the company is looking forward to embrace a comprehensive tunnel asset management approach including data management, asset condition evaluation and tunnel tailored best intervention strategy going from corrective to preventive maintenance approach.

References

1. Ministry of Infrastructure (2022) Linee guida per la classificazione e gestione del rischio, la valutazione della sicurezza ed il monitoraggio delle gallerie esistenti. D.M. 01/08/2022, n. 247
2. Mazzola M, Alessio C et al (2023) Tunnel Renewal Strategy: The cases of Castello 1 left, San Fermo right and Colle Marino left tunnels. In: Proceedings of the ITA-AITES World Tunnel Congress 2022, Athen, Greece
3. Alessio C, Baccolini L et al (2024) The 365 km tunnels assessment along ASPI Motorways Network – Key findings addressing risk analysis procedures and structural conditions evaluation and strategy of interventions. In: Proceedings of the ITA-AITES World Tunnel Congress 2024, Shenzhen, China
4. Frisiani S, Baccolini L et al (2024) Tunnel Renewal Strategy – The evaluation of the advantages through LCA. In: Proceedings of the ITA-AITES World Tunnel Congress 2024, Shenzhen, China
5. Alessio C, Poli A et al (2024) Colle Marino left tunnel renewal project: the first case-history of full lining reconstruction - from invert to crown - under difficult geotechnical conditions and future developments for working under traffic. In: Proceedings of the ITA-AITES World Tunnel Congress 2024, Shenzhen, China

6. Alessio C, Di Fiore D et al (2024) Development of a Tunnel Asset Management Tool from a Risk-Based Approach. In: Proceedings of the ITA-AITES World Tunnel Congress 2024, Shenzhen, China

Innovative Project Management for Sustainable Infrastructure: Coastal Highway Upgrade in Belize

Alessio Gori(✉) and Enea Sermasi

Politecnica Ingegneria ed Architettura, 220 Via Galilei, Modena, Italy
agori@politecnica.it

Abstract. The Coastal Highway Upgrade project in Belize represents a major advancement in sustainable infrastructure, transforming a 59-km stretch of flood-prone, unpaved road into a resilient, climate-adaptive highway. This project enhances connectivity, safety, and economic development in southern Belize by integrating innovative engineering solutions and environmental management. Key technical innovations include the use of geosynthetic materials for soil stabilization, advanced drainage systems to mitigate flooding, and elevated roadbeds to ensure continuous operation during extreme weather events. In addition, the project incorporated solar-powered streetlights and traffic management systems, reducing reliance on fossil fuels and aligning with Belize's renewable energy goals. Environmental considerations were central to the project, with comprehensive Environmental Impact Assessments (EIA) guiding the implementation of bioengineering techniques, sediment traps, and wildlife corridors to protect biodiversity. Social and economic sustainability were also prioritized, with the project creating jobs for local communities, offering training, and improving access to essential services. The project addressed key Environmental, Social, and Governance (ESG) challenges, balancing development with ecosystem protection and community engagement.

By adopting a multidisciplinary approach to design, construction, and management, the Coastal Highway Upgrade demonstrates how megaprojects can achieve sustainable development goals while mitigating environmental risks. This paper highlights the project's technical, environmental, and socio-economic contributions, providing valuable insights for future infrastructure projects in regions facing similar challenges, as well as the pivotal role of the adoption of the Building Information Modeling (BIM) in order to ensure completeness, reliability and compliance of the project, then consistency throughout the entire construction process.

Keywords: Sustainability · Stakeholder Engagement · ESG · BIM · Flood Management · Innovative Engineering

1 Introduction

The Coastal Highway Upgrade project in Belize is a landmark infrastructure initiative aimed at enhancing a vital 59-km section of unpaved and flood-prone road, transforming it into a climate-resilient highway (Fig. 1). This stretch connects the town of La

E. Favari et al. (Eds.): MeRIT 2024, LNCE 600, pp. 197–209, 2025.
https://doi.org/10.1007/978-3-031-92797-3_13

Democracia, located between Belize City and the National Capital City Belmopan, with Dangriga, forming a crucial link between the Country's economic and administrative heart and its southern regions (Fig. 2). Furthermore, two small villages are accessible only by this road, namely Gales Point and Mullins River.

Fig. 1. Road section of the former Coastal Highway on gravel and sand (ante-operam)

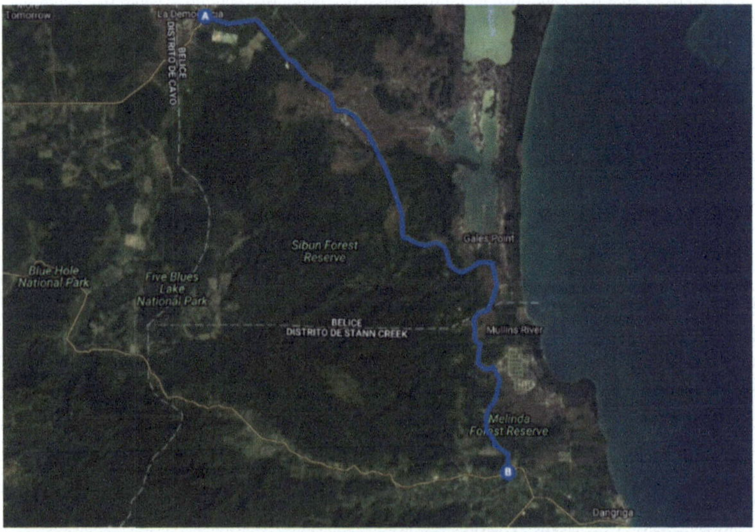

Fig. 2. Location of the Coastal Highway in Belize overlapped to an orthophoto

The road, previously vulnerable to extreme weather events (Fig. 3 and Fig. 4), has long been in need of a comprehensive upgrade to facilitate the safe movement of people and goods. By addressing the critical needs for improved connectivity, safety, and socio-economic growth, this project not only modernizes transportation in the region but also embodies forward-thinking sustainable design principles.

Fig. 3. Former Coastal Highway (location Sibun Area) flooded in October/November 2015, due–Tropical Cyclone Patricia

Fig. 4. Former Coastal Highway (location La Democracia) flooded in January 2018 just after a few hours of heavy rains

Throughout its development, the Coastal Highway Upgrade has embraced a holistic approach to sustainability, prioritizing environmental management, social equity, and the integration of renewable energy solutions. The project serves as a model for how infrastructure projects in vulnerable regions can simultaneously meet present demands and address future challenges posed by climate change. Thus, the project's impact on sustainable megaproject development can definitely be focused, at the same time, on technical innovations and engineering needs, as well as environmental impact mitigation, and the socio-economic benefits for the communities involved. Moreover, the use of Building Information Modeling (BIM) [3] as a critical project management tool will be highlighted to demonstrate how large-scale infrastructure projects can address modern challenges.

2 Project Management and ESG Implementation

The Coastal Highway Upgrade showcases how modern infrastructure projects can integrate Environmental, Social, and Governance (ESG) principles into their core. In an era when stakeholders are increasingly focused on the long-term environmental and social impacts of major developments, this project highlights the importance of aligning with both local and international ESG standards [9]. From the outset, the project's planning and design stages were informed by an in-depth analysis of regulatory frameworks, ensuring full compliance with Belizean environmental laws while also meeting the high expectations of global ESG benchmarks.

One of the standout elements of the project's management was its comprehensive stakeholder engagement strategy from the very beginning up to the works completion, a constant element in several years of project management (Fig. 5).

Fig. 5. A weekly stakeholders meeting and on-field visit during the construction phase

Recognizing the importance of community buy-in, the project team prioritized regular consultations with key local stakeholders, including community leaders, environmental groups, and government officials. This continuous dialogue ensured that local knowledge, particularly concerning the environmental and cultural landscape, was incorporated into the decision-making process. Importantly, it also provided a platform for addressing community concerns, ranging from the potential displacement of local wildlife to the social disruptions caused by construction activities. By adopting a transparent, inclusive approach, the project was able to build trust among the local population, ultimately contributing to its successful execution.

Risk management [7] was another critical aspect of the Coastal Highway Upgrade. A key challenge in infrastructure projects of this scale is the management of potential environmental and social risks. Early in the planning stages, the project team developed a robust risk management framework that identified potential hazards—ranging from ecosystem disruptions to community displacement—and implemented strategies to mitigate these risks. For example, the team worked closely with environmental scientists to assess the potential impacts of construction on local wildlife habitats. By taking proactive measures, such as the re-routing of the highway to avoid sensitive areas, the project minimized its environmental footprint.

3 Technical Innovations in Design and Construction

One of the most striking aspects of the Coastal Highway Upgrade is its array of technical innovations, which have improved the road's resilience to extreme weather conditions. At the heart of these innovations is the comprehensive use of BIM, which helped mitigate risks by identifying potential issues early in the design phase, thus reducing costly changes during construction.

Flood management was a primary concern during the highway's design phase. For that purpose, specific and innovative surveys campaign have been set: beyond the traditional ones, like topographic, cadastral, traffic and geotechnical investigations, the adoption of LIDAR (Fig. 6) [10] and bathymetric surveys has allowed the design to be based on a digital twin of the actual terrain morphology.

Fig. 6. LIDAR equipment and instruments for 3D-terrain surveys and modeling

Traditional road construction in flood-prone areas often requires frequent maintenance, as the combination of heavy rainfall and soft soil can lead to rapid degradation. However, the Coastal Highway has achieved enhanced durability, offering better resistance to soil erosion and flooding [6]. This not only reduces the frequency of repairs but also contributes to the highway's long-term sustainability, ensuring that it can withstand Belize's often unpredictable weather patterns (Fig. 7).

Fig. 7. Manatee Bridge under construction (drone image)

The coastal regions of Belize are particularly vulnerable to hurricanes, tropical storms, and the heavy rains that accompany these events. To mitigate the risk of flooding [5], the project included the installation of a comprehensive stormwater management system. Culverts, retention basins, and strategically placed drainage ditches were incorporated along the entire length of the highway, ensuring that stormwater is efficiently diverted away from the road surface. This not only protects the structural integrity of the highway but also safeguards the surrounding environment, preventing soil erosion and reducing the likelihood of water contamination in nearby ecosystems.

In addition, the project employed elevated roadbeds in sections of the highway that are particularly susceptible to flooding. These raised roadbeds, combined with the drainage systems, ensure that the highway remains operational even during extreme weather events. This level of foresight in design and construction has made the Coastal Highway one of the most resilient infrastructure projects in the region, setting a new standard for climate-adaptive infrastructure.

4 BIM Integration and Its Role in Project Success

A critical component of the project's success was the use of Building Information Modeling (BIM) (Fig. 8). This methodology provided a unified platform for collaboration between architects, engineers, and contractors.

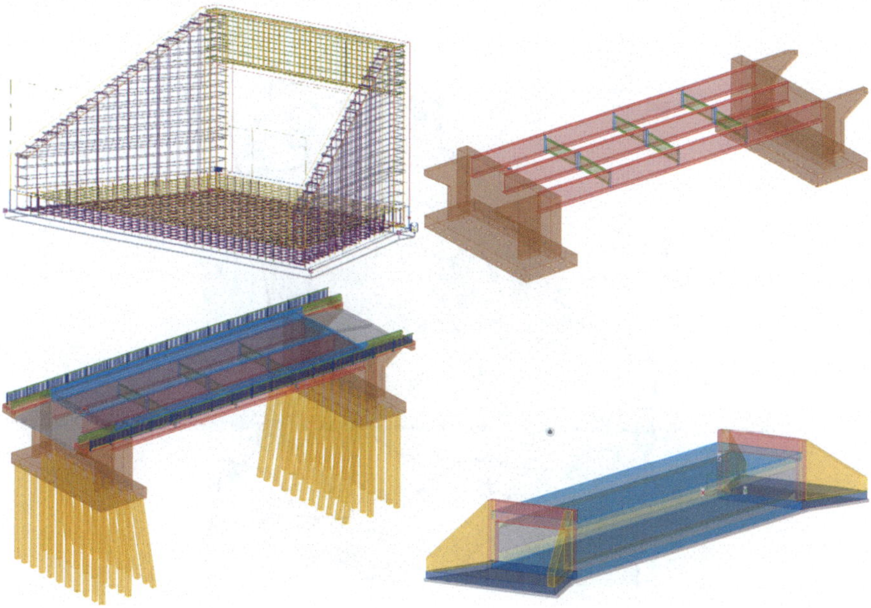

Fig. 8. Structural design of bridges and hydraulic culvert by adopting BIM software

By using BIM the project team was able to simulate various design options, optimize the allocation of resources, and foresee potential clashes between different systems. This enhanced collaboration between stakeholders ensured that the project was delivered on time and within budget.

Moreover, BIM facilitated better construction management by enabling detailed scheduling and sequencing of construction activities. By visualizing each stage of construction, the project team could predict challenges and adjust timelines proactively. The integration of BIM into the Coastal Highway Upgrade not only improved efficiency but also reduced waste, contributing to the project's overall sustainability goals.

The ability to maintain real-time data on construction progress, material usage, and environmental conditions via BIM provided a robust data-driven approach to managing the project. This digital transformation ensured transparency and accountability at all levels, leading to a more streamlined workflow and better communication across teams.

5 Environmental Impact Assessment and Mitigation Strategies

The Coastal Highway Upgrade was guided by a thorough Environmental Impact Assessment (EIA), which played a central role in shaping the project's approach to environmental sustainability. The EIA process involved an exhaustive analysis of the project's potential effects on local ecosystems, including impacts on wildlife, water resources, and soil integrity. Key risks identified through this process included habitat fragmentation, the potential for water contamination due to runoff, and the erosion of soil in sensitive areas. Each of these risks was addressed through carefully crafted mitigation strategies aimed at minimizing long-term environmental damage (Fig. 9).

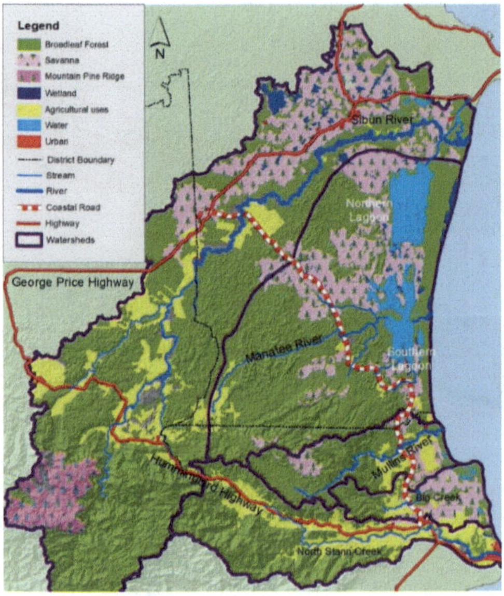

Fig. 9. Land uses along the Coastal Highway

To combat soil erosion and stabilize embankments, the project incorporated bioengineering techniques that involved planting native vegetation along the highway's slopes. These plants not only help anchor the soil, preventing landslides and erosion, but also contribute to the restoration of local ecosystems by providing new habitats for wildlife. This approach demonstrates the project's commitment to environmental restoration, as it goes beyond simply preventing damage to actively enhancing the ecological value of the surrounding landscape.

In terms of water management, the construction of sediment traps and retention basins along the highway helped capture stormwater runoff, preventing it from polluting nearby rivers and wetlands. These water management strategies ensured that the highway did not disrupt the natural flow of water in the region, maintaining water quality and preserving aquatic ecosystems. Additionally, strict waste management protocols were implemented during the construction phase. By recycling construction materials and properly disposing of hazardous waste, the project minimized its environmental footprint and adhered to international environmental standards.

6 Renewable Energy Integration

The Coastal Highway Upgrade's commitment to sustainability is perhaps best exemplified by its integration of renewable energy solutions. One of the key innovations was the installation of solar-powered streetlights along the highway. These lights provide continuous illumination even in areas that are not connected to the national power grid, enhancing the safety and security of the highway. In addition to the streetlights, solar energy was also integrated into the highway's traffic management system. Solar-powered sensors and monitoring devices were installed along the route to collect real-time data on traffic conditions, road surfaces, and weather patterns. This information allows for better decision-making regarding road maintenance and traffic control, improving the efficiency and safety of the highway while reducing its carbon footprint.

The project also made use of renewable energy during the construction phase. Several of the construction vehicles and machinery were powered by biofuels, significantly reducing the project's reliance on fossil fuels. This focus on renewable energy not only lowered the project's environmental impact but also set a new precedent for future infrastructure developments in Belize and beyond.

7 Social and Economic Sustainability

Beyond its environmental benefits, the Coastal Highway Upgrade has had a transformative impact on the social and economic landscape of southern Belize. By significantly reducing travel time—up to 40% in some cases—the highway has improved access to key economic hubs, facilitating trade, tourism, and agricultural activities. The region, which has long suffered from inadequate infrastructure, is now better positioned to tap into new economic opportunities, creating a ripple effect of economic development that is expected to last for decades.

From a social perspective, the project has prioritized inclusivity and community engagement. Throughout the planning and construction phases, the project team worked closely with local communities to ensure that the highway met their needs. Public consultations were held regularly, providing residents with a forum to voice their concerns and offer feedback. This participatory approach not only helped to ensure that the highway was built in a way that minimized disruption to daily life but also fostered a sense of ownership and pride among local residents.

The project's economic impact extends to job creation as well. Hundreds of local workers were employed during the construction phase, many of whom received specialized training in construction techniques, environmental management, and safety protocols. This investment in human capital has left a lasting legacy, as the skills acquired by the workforce will be valuable for future infrastructure projects in the region. The Coastal Highway Upgrade's contributions to Sustainable Development Goals (SDGs) 8 (Decent Work and Economic Growth) and 10 (Reduced Inequalities) [1, 2] have been widely recognized, including the project's receipt of the SDG Pioneer Italy 2024 Award.

Despite its overall success, the Coastal Highway Upgrade project encountered several notable challenges, particularly in balancing the competing demands of infrastructure development and environmental protection [4]. One of the primary difficulties involved the preservation of delicate ecosystems, many of which were located in areas crucial to the highway's expansion. The project team had to navigate the complex task of expanding the road network without causing undue harm to these sensitive environments. By adopting a multidisciplinary approach, which involved the collaboration of environmental scientists, civil engineers, and local stakeholders, the team was able to minimize environmental damage while still achieving the necessary improvements in infrastructure.

The project also faced significant challenges in managing local community expectations. Many residents initially expressed skepticism regarding the project, concerned about how construction activities might disrupt their lives or potentially harm the surrounding environment. To address these concerns, the project team implemented an extensive communication strategy that emphasized transparency and regular updates. Community outreach meetings were held frequently, allowing residents to stay informed about the project's progress and offering them a platform to express their concerns. This open dialogue not only built trust between the project team and the local population but also helped to foster a sense of ownership among residents, who began to see the highway as a community asset rather than an external imposition.

Another challenge arose in implementing the Environmental, Social, and Governance (ESG) standards that were integral to the project's design. While the project successfully adhered to international ESG frameworks, maintaining compliance with these standards throughout the various phases of the project required continuous oversight and coordination. This was particularly true during the construction phase, where on-the-ground conditions often required quick adjustments to stay aligned with the project's environmental and social goals. Despite these challenges, the project was able to achieve its objectives without significant delays or cost overruns, demonstrating the effectiveness of a well-structured ESG strategy in large-scale infrastructure development.

8 Water and Waste Management Solutions

Given the region's susceptibility to flooding, effective water management was a critical component of the Coastal Highway Upgrade. The project included a state-of-the-art stormwater management system designed to control runoff during heavy rainfall, on the basis of various scenarios related to specific design return period of the events and then associated to the hydraulic risk along the new infrastructure (Fig. 11). Culverts, drainage ditches, and detention basins were placed strategically along the highway to prevent water from pooling on the road surface and causing damage. These systems not only protect the structural integrity of the highway but also reduce the risk of flooding in surrounding communities, making the project a model for climate-adaptive infrastructure development (Fig. 10).

In terms of waste management, the project adhered to strict environmental standards throughout the construction phase. Recyclable materials were repurposed wherever possible, and hazardous waste was disposed of according to both local regulations and international best practices. By prioritizing both water and waste management, the Coastal

Fig. 10. Hydraulic protection measures under construction (hydraulic gabions) along the new road (drone image)

Highway Upgrade set a new benchmark for sustainable infrastructure development in Belize, demonstrating that large-scale projects can be carried out without compromising environmental integrity.

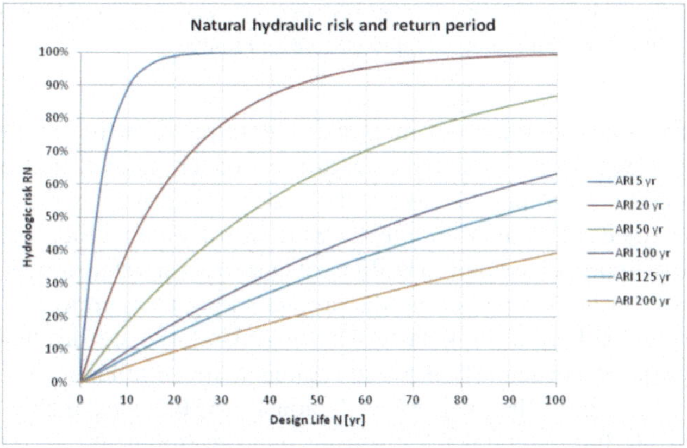

Fig. 11. Natural hydraulic risk and design return period

9 Conclusions

The Coastal Highway Upgrade project in Belize has set a new precedent for sustainable infrastructure development, proving that it is possible to undertake large-scale projects while maintaining a strong commitment to environmental, social, economic sustainability. By embracing innovative design and construction techniques, prioritizing environmental mitigation efforts, incorporating renewable energy solutions, the project serves as a model for megaprojects [8], particularly those in regions vulnerable to climate change (Fig. 12).

Fig. 12. Two set of couple image: comparisons between the existing Coastal Highway (top image, ante-operam) and the new road almost completed (bottom image)

The project's success lies not only in its ability to improve the region's transportation infrastructure but also in its capacity to address broader societal and environmental challenges. Through its comprehensive stakeholder engagement strategy, the project ensured that the voices of local communities were heard and that their needs were met. This collaborative approach helped to build trust and foster a sense of ownership among local populations, further strengthening the project's long-term impact.

The lessons learned from the Coastal Highway Upgrade are invaluable for future infrastructure projects, particularly those in regions facing similar socio-economic and environmental challenges. The project has demonstrated that sustainable infrastructure is not only feasible but essential for ensuring that development can proceed in harmony with environmental stewardship and social responsibility. As the need for climate-resilient infrastructure grows, the experiences and insights gained from the Coastal Highway Upgrade will serve as a blueprint for similar projects, guiding them toward more sustainable and inclusive outcomes.

In conclusion, the Coastal Highway Upgrade is a testament to the power of innovation, collaboration, and sustainability in shaping the future of infrastructure development. By prioritizing the needs of local communities, protecting the environment, and leveraging cutting-edge technologies, the project has made significant contributions to both local development goals and global sustainability targets. As the world continues to grapple with the challenges posed by climate change and population growth, projects like the Coastal Highway Upgrade offer a vision of how sustainable infrastructure can contribute to a more resilient and equitable future.

References

1. Brundtland GH (1987) Our Common Future. Oxford University Press, Oxford
2. United Nations (2015) Transforming our world: the 2030 Agenda for Sustainable Development. UN Publishing, New York
3. Eastman C, Teicholz P, Sacks R, Liston K (2011) BIM Handbook: A Guide to Building Information Modeling for Owners, Designers, Engineers, Contractors, and Facility Managers, 2nd edn. Wiley, Hoboken
4. Rittel HWJ, Webber MM (1973) Dilemmas in a general theory of planning. Policy Sci 4(2):155–169
5. Hegger DLT et al (2016) Toward more flood resilience: Is a diversification of flood risk management strategies the way forward? Ecol Soc 21(4):52
6. Wilkinson SJ, Remøy H, Langston CA (2014) Sustainable Building Adaptation: Innovations in Decision-Making, 2nd edn. Wiley-Blackwell, Oxford
7. Aven T (2016) Risk assessment and risk management: review of recent advances on their foundation. Eur J Oper Res 253(1):1–13
8. McKinsey & Company: Bridging Global Infrastructure Gaps. https://www.mckinsey.com/business-functions/operations/our-insights/bridging-global-infrastructure-gaps. Accessed 11 Nov 2024
9. World Bank (2017) Environmental and social framework: setting environmental and social standards for investment project financing. The World Bank, Washington D.C.
10. Rezaei M, Sardroud JM, Alavi M (2020) Integration of BIM and GIS in sustainable infrastructure: a systematic literature review. In: Proceedings of the 37th International Symposium on Automation and Robotics in Construction (ISARC 2020), pp 121–128. IAARC, Kitakyushu

Transforming Megaprojects: Insights from High-Speed Rail Development

Jithin Arakkatt Shaji[1(✉)] and Ritwik Aravindakshan[2]

[1] Vinci Construction, HS2 Project, Watford, UK
jithinashaji@gmail.com

[2] Department of Mechanical Engineering, Toc H Institute of Science and Technology, Kochi, India

Abstract. Projects that bring about substantial and enduring changes to society and the economy are referred to as transforming megaprojects. These projects often span over decades, encompassing an entire economic cycle, governmental shifts, periods of growth and recession, multiple five-year plans, financial budgets, political shifts, social unrest, and unforeseen events such as pandemics and natural disasters. Given this dynamic context over such a prolonged timeline, initial prospects and objectives are subject to significant changes as new challenges arise and opportunities emerge. This paper examines the factors that influence the shifting landscape of megaprojects and their core objectives with a case study.

Keywords: Transforming · megaproject · overruns · changing landscape · challenges · and opportunities

1 Introduction

Megaprojects are large-scale, complex projects that typically cost more than one billion US$, involve multiple public and private stakeholders, and take many years to build by transforming the lives of millions of people [1]. It is estimated that the world needs to spend about $57 trillion on infrastructure and built environments by 2030 to enable the anticipated levels of GDP growth globally [2]. While acknowledging the potential for upward revisions, a cautious estimate places the global megaproject market within an annual expenditure range of US$6 and US$9 trillion [1]. This in turn fosters significant job creation, and technological advancements, stimulating economic growth with long-term sustainability. However, despite their ambitious objectives, these projects frequently experience significant changes in scope and prospects during the execution phase. A study examining a sample of 3022 projects revealed that only 27% were completed within or under budget, while the remaining 73% encountered budget overruns, schedule delays, or significant deviations from their original scope and objectives [3]. From the ongoing projects such as the High-Speed 2 (HS2) railway in the United Kingdom, High-Speed Rail (HSR) development in the United States, and the NEOM Line in Saudi Arabia, to the design and planning of Australia's HSR system, megaprojects are dynamic and prone to substantial modifications over the time. Global ventures like the Panama

E. Favari et al. (Eds.): MeRIT 2024, LNCE 600, pp. 210–227, 2025.
https://doi.org/10.1007/978-3-031-92797-3_14

Canal expansion and the Channel Tunnel exemplify how changes in scope, budget, and timelines are common and have significant and far-reaching consequences up until handing over to the end users.

The evolution of mega projects is often influenced by changes to the technological and regulatory environments in which they operate. As technology advances and amendments to regulations occur, projects adapt to meet the revised standards, absorb innovative solutions, and mitigate unforeseen risks. These adaptations can impose re-engineering or redesign efforts, which in turn impact cost and schedule estimates. The reliance on experimental technologies or procedures can also introduce uncertainties, making it challenging to maintain the original scope. The primary objective of this study is to analyze the key factors influencing the success of megaprojects through a case study and to conclude with observations for integrating these factors into the planning and execution of future megaprojects.

2 Background

2.1 Global Overview of Ongoing Megaprojects

The exponential growth of the global population has resulted in massive urbanization and energy needs to meet the requirements and aspirations of society. Projects such as roads, railways, metros, tunnels, dams, airports, mobility hubs, power plants and self-contained cities are required to provide infrastructure to the increasing inhabitants. While these massive developments scale up, their environmental impacts are significant, resulting in an increasing imperative to plan more sustainable solutions for such large-scale developments. Globalization is the driving factor for the numerous megaprojects progressing around the world with varied objectives and challenges. Megaprojects are derived from the urban entrepreneurial strategy to attract investment to the cities as a solution to globalization challenges [4]. Therefore, megaprojects are on the agendas of policymakers across the globe.

The growth of developing countries has significantly increased energy use and emissions. The United Nations estimates that the world urban population will be nearly 6.4 billion by 2050 [5]. It is evident from the global snapshot of the mega projects that the geographic distribution highlights a disproportionate focus on developing and oil-exporting economies (see Fig. 1). Many multibillion-dollar projects are being developed in the Middle East aligning the Vision 2030 plans that the oil exporting countries like Saudi Arabia, Qatar, and United Arab Emirates (UAE) have in place. The Vision 2030 plan aims to create a thriving post-hydrocarbon economy for the area by creating tourist attractions, entertainment venues, and financial hubs with self-sufficient hi-tech cities [6]. Before the end of the decade, the world will see the first construction megaproject with a total cost estimation exceeding one trillion US$ [7].

Figure 2 illustrates some major megaprojects with budgets over 25 billion US$. Here is a brief overview of the top five projects on the list [9].

The TEN-T or Trans-European Transport Network is a wide network of roads, railway lines, inland waterways, maritime shipping routes, ports, airports, and railroad terminals around Europe. The estimated total cost is $600 billion with the prime objective

Fig. 1. A global snapshot of megaprojects in progress [8–12].

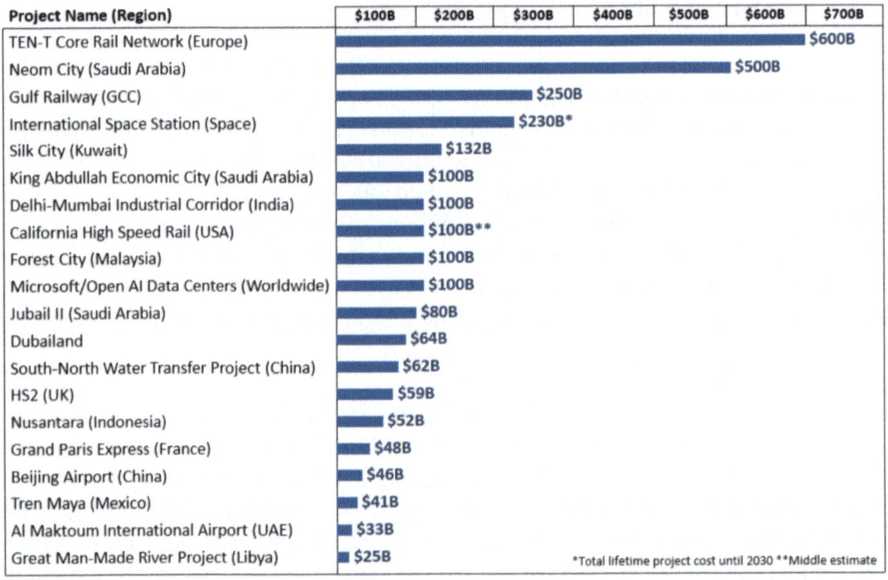

Fig. 2. Estimated cost of selected megaprojects currently under construction globally with a value greater than $25 billion [8, 9].

to develop a sustainable infrastructure network by reducing the environmental impact of transport while increasing energy efficiency and safety.

The NEOM is a self-contained futuristic city with a planned area of 26,500 Sqm being built in Tabuk Province of Saudi Arabia. The estimated cost is $500 billion however future developments may sum up to 1.5 trillion US$.

The Gulf Railway is a rail system to connect all six Gulf countries with a total length of 2177 km and the project is estimated to cost $250 billion. The project is expected to provide free movement within the GCC and boost intra-GCC trade by providing unhindered rail transport from Kuwait to Oman.

In April 2024, a joint venture was signed between Oman and the United Arab Emirates to deliver the East-West rail wing spanning 303 Km within the Gulf Railway network connecting Abu Dhabi to the Omani port city of Sohar.

The International Space Station is an extensive space station being built in the low Earth orbit to perform microgravity and space environment experiments. The project is in collaboration with five space agencies: NASA (US), Roscosmos (Russia), ESA (Europe), JAXA (Japan) and CSA (Canada) with an estimated cost of $230 billion as a total lifetime project cost until 2030.

The Silk City also known as Madinat Al-Hareer is a mix of infrastructure projects developed in phases in northern Kuwait. The project with an estimated cost of $132 billion is part of Kuwait Vision 2035 to create a Northern Economic Zone for developing beneficial economic ties with Iraq and Iran. The intended life span of ISS has been extended several times. Most of the station is modular and with wearing out of the parts the new part can be replaced or augment the original. The station will continue to be a working laboratory and outpost in orbit until 2030. It is deemed an unprecedented achievement in global human endeavours to build and utilize a research platform in space. Since 2000, it has evolved as a highly potential microgravity laboratory and in its third decade of continuous human presence, the station is a significant pod for demonstrations and scientific investigations in a wide spectrum of space.

2.2 Prospects and Barriers

A hallmark of mega project evolution is scope creep, the gradual expansion of a project's objectives, deliverables, or features beyond the initial plan. Scope creep can be attributed to various factors, including stakeholder demands, regulatory changes, and the identification of additional requirements during project execution. While some modifications may enhance the project's value, may result in cost overruns, delays, and, in certain cases, the abandonment of key components. For example, in the HS2 project, escalating costs and shifting political priorities led to the cancellation of the northern sections originally intended to connect Manchester and Leeds to the high-speed rail network from London and Birmingham. This change substantially altered the project's economic benefits and long-term prospects, raising questions about its ability to fulfil its initial promises of national connectivity and regional economic development.

Figure 3 illustrates the magnitude of cost overruns on various projects across the globe. The infrastructure developments are more susceptible to cost overruns due to scope changes, design developments, geological conditions and changing stakeholder expectations. The multifaceted reasons behind the evolving scope and outcomes of mega projects often include overly optimistic projections of both costs and benefits. Decision-makers, under the influence of political and public pressure, may underestimate the complexity and risks associated with such endeavours, leading to unrealistic timelines, budget estimates, and project deliverables.

As projects progress, unforeseen challenges such as technical difficulties, environmental concerns, stakeholder opposition, and economic fluctuations frequently necessitate modifications to the original plan. For instance, shifts in global market conditions,

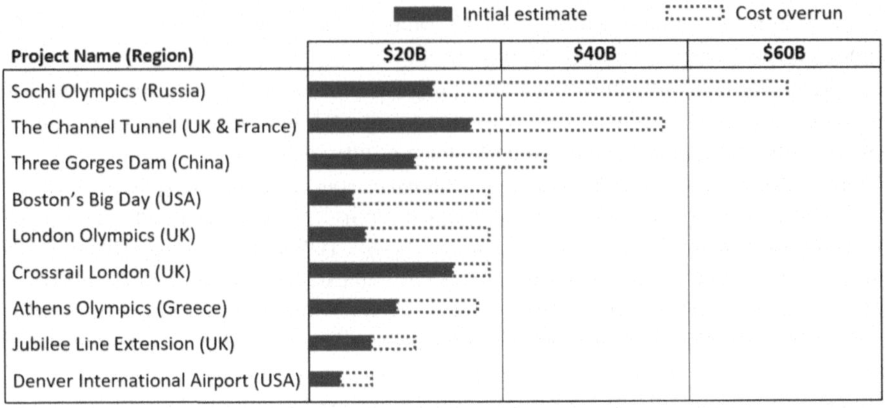

Fig. 3. Major projects that went over the budget, converted to USD and adjusted for inflation [13].

such as increased material costs or inflation, can significantly alter the financial landscape of a mega project. Additionally, political changes and evolving public expectations may require adjustments to the project's scope to align with emerging priorities.

The Neom project, a $500 billion megacity initiative in Saudi Arabia, originally envisioned accommodating 1.5 million inhabitants by 2030. However, contemporary reports suggest a recalibration of these aspirations within the broader context of the Kingdom's Vision 2030 program. This recalibration is indicative of a potential reduction of inhabitants to approximately 300,000 residents by 2030. This adjustment follows a phased construction approach, with only a 24 km segment of the envisioned 170 km stretch expected to be completed by 2030 [14]. Originally conceptualized as a technologically advanced urban hub that aims to diversify the economy away from oil. The project is conceptualized to be powered only by renewable energy incorporating cutting-edge transportation systems like aerial vehicles and automated services. The rescaling of the Line highlights the challenges with complexities associated with such sophisticated projects however overall vision for Neom remains intact, the reality of implementing such a transformative endeavour is likely to involve a more gradual and phased approach.

The high-speed rail project in California is facing severe scope creep and budget overruns, the project encompasses over 800 miles of rail, with around 24 stations. The project is currently in Phase 1 and extends between San Francisco to Los Angeles and Anaheim. With issues over land acquisition, inflation, local opposition, environmental lawsuits, and changes to scope the project cost forecast is four times the initial budget. Figure 4 summarizes the change in project cost over the last two decades.

Furthermore, the governance and project management structures employed in mega projects play a pivotal role in shaping their evolution. Factors like deficient planning, ineffective risk management, and poor communication among stakeholders are common contributors to scope changes. Megaprojects involve a multitude of stakeholders, including governments, private contractors, global supply chain and logistics, financiers, and local communities, each with distinct interests and priorities. These competing interests often lead to conflicts, delays, and compromises that alter the project's trajectory.

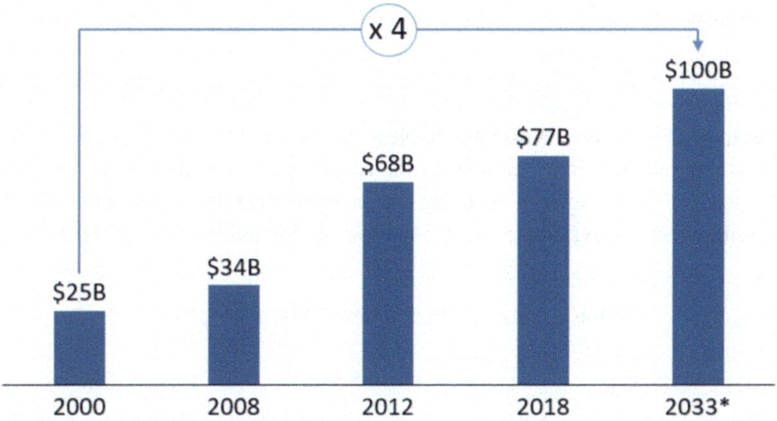

Fig. 4. Evolution of total budget estimation for California High-Speed Rail Project, *Highest projected amount by 2033 completion date [16].

3 Methodology

This research employs a qualitative inductive methodology, following an extensive literature review, and a series of interviews with key project stakeholders of major megaprojects across the globe. The primary objective of this paper is to investigate the reasons behind the frequent changes in scope that occur during the execution phase of mega projects, the resulting consequences, and the strategies that can be implemented to manage these changes effectively. By examining a diverse range of case studies, interacting with the key delivery personnel, and analyzing the underlying causes of scope shifts, this paper aims to provide valuable insights into how such projects can navigate the complexities of execution to achieve their desired outcomes. Ultimately, comprehending the dynamics of scope change in mega projects is essential for refining project management practices and ensuring the successful delivery of large-scale infrastructure and development initiatives.

3.1 Literature Review

A semi-structured literature review was initially conducted and comprehensively explored the available data. Various keywords like “megaprojects”, “transformational projects”, “overruns”, “changing landscape”, “challenges”, “scope creep”, and “opportunities” were used to explore the body of knowledge and get diverse perspectives within the field. The search was extended to legislative changes, policy settings, master plan developments, fiscal challenges, and annual budgets. Following an initial review of abstracts, relevant papers were selected for in-depth analysis. Further to elucidate the transformational dynamics of a megaproject in navigating the barriers encountered during its journey from evolution to implementation, we have examined projects such as the NEOM Line in Saudi Arabia, high-speed rail projects in California, Texas, and the Northeast Corridor of the United States, and the HS2 in the United Kingdom.

3.2 Interviews

A set of semi-structured interviews were conducted with construction professionals working in megaprojects across the globe (UK, France, USA, Middle East, India, Philippines, Maldives, Singapore, Australia & New Zealand). The interviewees were within the professional and personal network of the researchers. The interviews were focused on identifying the challenges in meeting the project objectives, budget, and schedule. Table 1 summarizes the type of project, location, and profiles of the interviewees.

Table 1. Interviewee Profiles and Demographics

No.	Type of Project	Location	Sector	Designation
1	Power Plant	UK	Contractor	Project Manager
2	Railway	France	Contractor	Construction Manager
3	Railway	USA	Designer	Design Manager
4	Megacity	Middle East	Client	Construction Director
5	Hospital	India	Consultant	Project Director
6	Railway	Philippines	Consultant	Construction Director
7	Airport	Maldives	Consultant	Planning Engineer
8	Retail Building	Singapore	Contractor	Project Coordinator
9	Water supply	Australia	Client	Associate Director
10	Roads	New Zealand	Contractor	Section Engineer

3.3 Case Study

The rescoping of HS2 Phase 2 was considered as the case study as it represents the latest update and one of the author's firsthand experiences with the project. This project provides a valuable perspective to analyze the evolution of a large-scale initiative, with a particular focus on cost-benefit assessment. In October 2023, the Prime Minister of the United Kingdom announced that "…the facts around HS2 have changed. Costs have more than doubled. It has been repeatedly delayed. The pandemic has significantly changed the travel patterns it was originally designed to serve. In short, HS2 traps enormous sums of money that could be far better spent elsewhere [17]." This exemplifies a quintessential example of project goal reassessment necessitated by exogenic influences.

Case Description

HS2 is a nascent High-Speed railway line initially conceived to link London to the major cities in the North of the United Kingdom, such as Birmingham, Manchester, and Leeds. The project was proposed to stimulate economic growth by improving the connectivity between major cities and to reduce the imbalance between the North and South of England. The project was proposed in 2009 by the Department of Transport to address the growing need for efficient transportation between the north and south of England by expanding rail connectivity and improving access to the capital city from northern urban centers. A future extension was envisioned up to Scotland to have the full potential of the intended transportation corridor. The proposed HS2 route into London will pass within proximity of the existing HS1 line which is a 108 Km link between London and the Channel Tunnel to France, creating a unique opportunity to connect the two high-speed rail networks. The Department for Transport (DfT) initially envisioned a direct link between HS2 and HS1, which would have enabled HS2 trains from the North of England to bypass London Euston and connect directly to the European rail network. This connection would have realized the long-standing ambition of the Regional Eurostar scheme, first proposed in the 1980s, to provide direct rail services between the United Kingdom and continental Europe.

The HS2 railway project was initially conceptualized as an ambitious two-phase development to revolutionize Britain's rail infrastructure. The initial phase encompassed the construction of the London-to-Birmingham route, while the second phase extended the network to Manchester and Leeds (Fig. 5). Construction activities for phase one commenced in 2017, with completion anticipated by 2030 [18]. The subsequent phase two was strategically scheduled to begin in 2027 [19]. A pivotal milestone in the project's trajectory was achieved in November 2016 when the government provided its official endorsement, confirming the planned route and signaling a significant advancement in the project's implementation [20].

Fig. 5. The HS2 infrastructure map, including IRP amendments, published 15 Jun 2022 [18].

The project has undergone substantial scope reductions in recent years, fundamentally altering its original vision. The proposed route to Leeds was discontinued in November 2021, followed by the cancellation of the Golbourne connection to the West Coast Main Line in June 2022. With a significant development in October 2023, the planned route to Manchester was also terminated. Due to these sequential reductions, the project now consists solely of the London-to-Birmingham segment, along with a spur to the West Coast Main Line near Lichfield [21–23].

In response to these developments, in October 2023, the Chairman of the UK's National Infrastructure Commission, urged a pause on the sale or de-safeguarding of land designated for the recently abandoned HS2 routes, advocating for a period of reconsideration and exploration of alternatives. He stressed the importance of developing a well-conceived and integrated plan. In an interview with the BBC, the Chairman remarked, "I believe it is a mistake". The land should be retained for at least two to three years to allow for a reassessment of potential solutions within that space, providing an opportunity to identify a more cost-effective approach, rather than dismissing it prematurely [24].

A comprehensive analysis conducted in 2016 revealed a striking disparity: the cost-per-mile of HS2 Phase 1 was five times that of comparable European projects [26]. Subsequent studies have indicated even more significant cost differentials, with the Paris-Strasbourg high-speed link demonstrating remarkable efficiency by costing approximately twelve times less per mile [27]. When the critical decision to proceed with the full project was made in 2020, the cost of Phase 1 had more than doubled from its initial estimate in 2012. This substantial increase was driven by a confluence of factors, including overly optimistic cost projections, significant changes in project scope and timeline, and suboptimal delivery performance. The situation was further exacerbated by the unprecedented impact of the COVID-19 pandemic and elevated inflation rates in subsequent years, resulting in Phase 1 alone exceeding the original estimated cost for the entire HS2 project [27].

As illustrated in Fig. 6, the forecast cost of completion in 2023 for Phase 1 is three times the initial estimates from 2012. Further, the project's declining cost-benefit ratio emerged as a significant factor influencing the decision to revise the network's scope. As project costs have escalated, the relative benefits have diminished considerably (see Fig. 7). Initial assessments had projected an optimistic economic return of £2.30 for every £1 invested. However, recent forecasts indicate a substantially reduced return of only 80 pence per £1 of taxpayer investment [28]. This strategic decision to partially abandon the project reflects a calculated response to multiple factors, including escalating costs, declining cost benefits, the profound impact of the pandemic on travel patterns, and the inherent complexities of managing large-scale infrastructure projects in an increasingly volatile economic environment. While the London-to-Birmingham segment remains intact, the abandonment of the northern routes has raised significant questions about the project's ability to achieve its original objectives of rebalancing the UK economy and enhancing connectivity between northern and southern regions.

In a significant policy shift announced in October 2023, UK Prime Minister Rishi Sunak introduced the Network North initiative as a strategic replacement for HS2 Phase 2. In his announcement, he emphasized, "... every pound that we save from not proceeding

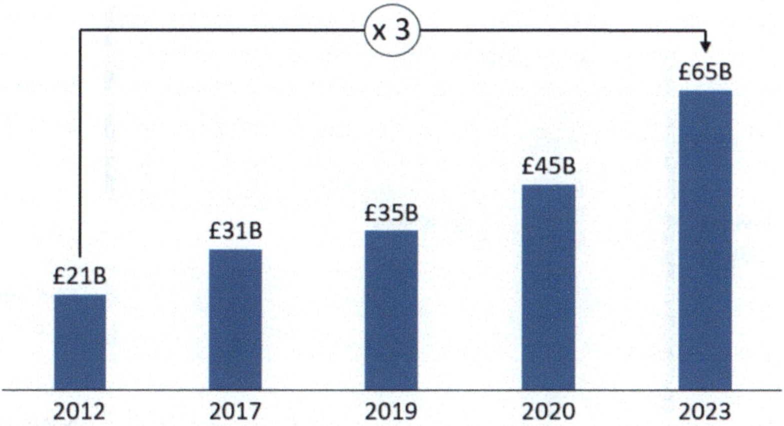

Fig. 6. The increasing costs of HS2 Phase 1 [25].

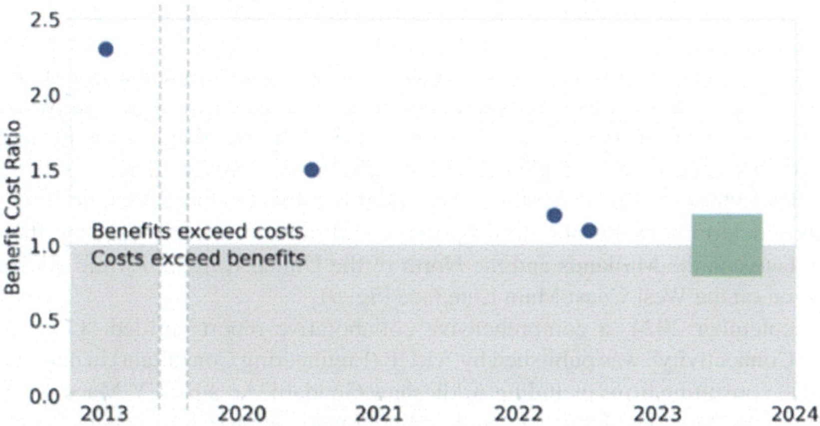

Fig. 7. HS2 assessed project benefits benefit analysis per £ spent [28].

with further phases of the scheme will instead be reinvested in hundreds of transport projects across the country, benefitting far more people, in far more places, far more quickly" [17]. This new concept focuses on developing multiple transportation networks designed to serve millions of individuals, in contrast to a single high-speed rail line that would primarily benefit a more limited population.

The rescoping of the project is expected to release £36 billion, which the government plans to strategically reinvest in infrastructure projects aimed at benefiting towns, cities, and rural areas across the United Kingdom. As detailed in Fig. 8, all funds will be systematically directed towards the communities that were originally set to benefit from the scheme. Of the total amount, £19.8 billion previously allocated to the northern leg of HS2 will be reinvested in projects across the North, £9.6 billion initially earmarked for the Midlands leg will be redirected to the Midlands, and £6.5 billion saved from the Euston Station development will be distributed across other regions of the country.

Fig. 8. HS2 Re-prioritized spending [29].

Following the October 2023 announcement, in March 2024, the mayors of Greater Manchester and the West Midlands collaboratively advocated for a new railway link. This proposed connection between Handsacre (HS2's Birmingham endpoint) and the Northern Powerhouse Rail at High Legh in Manchester was presented as a viable substitute for HS2 Phase 2 [30]. The initiative thoroughly evaluated three possible strategies, ultimately determining that the third option would most effectively enhance the connection between the Midlands and the North of the United Kingdom while alleviating congestion on the West Coast Main Line (see Fig. 9).

In September 2024, a comprehensive collaborative report entitled "Opportunity through Connectivity" was published by ARUP (Engineering Consultant) in conjunction with other consulting firms including Addleshaw Goddard, Arcadis, EY, Mace, Skanska, and Dragados. This detailed report outlined a strategic approach to enhance transport infrastructure between the Midlands and Northwest of the UK, emphasizing strengthening rail infrastructure to promote growth and address congestion on the West Coast Main Line (WCML) following the cancellation of HS2 Phase 2 [33]. The MNWRL outlined in the report detailed several potential design solutions making use of the same route as the former HS2 Phase 2, while adopting significant changes to lower the overall capital cost [33]:

A reduced design speed of 300 km per hour, as opposed to the initially proposed HS2 speed of 340–400 km per hour, while still adhering to the definition of high-speed rail and significantly surpassing the current maximum speeds of 120–175 km per hour for non-tilting trains on the WCML.

A smaller cross-section aligned with British rather than European specifications, necessitating less extensive construction work to circumvent obstacles and ensure seamless compatibility with the existing and future national rail network.

The replacement of HS2-specified concrete slab track with ballasted track, noting that concrete slab track has around 70% higher capital cost compared to ballasted track.

Fig. 9. The new line between Handsacre and Manchester Airport proposed to replace HS2 Phase 2 published on 21-Mar-2024 [32].

A simplified rail systems integration with the existing Network Rail infrastructure, particularly in the vicinity of Crewe.

The proposed initiative advocated for a phased approach to realizing its broader objectives. The MNWRL is envisioned to be implemented in two distinct stages:

Initial Phase: The Staffordshire Connector, encompassing the segment between Lichfield and Crewe, along with necessary enhancements to the existing Northern Powerhouse Rail, is planned to proceed based on the authorization granted by the High-Speed Rail Act 2021.
Subsequent Phase: This will focus on the design and planning of the Cheshire Connector, comprising the northern segment connecting Liverpool and Manchester, in conjunction with new railway lines to the Northern Powerhouse Rail.

A sophisticated blended funding strategy was proposed, drawing on case studies of comparable models across Europe and other parts of the world. This approach aims to enhance affordability, improve cost management, mitigate risks, and support timely project delivery. The commercial framework envisions the creation of a privately led Special Purpose Vehicle (SPV) for the MNWRL, operating under a concession agreement with the government, as illustrated in Fig. 10. This agreement is structured to balance public sector oversight and governance with the private sector's ability to deliver services efficiently. The SPV will employ a mix of debt and equity financing over the project's duration, with revenue streams derived from sources such as track access charges.

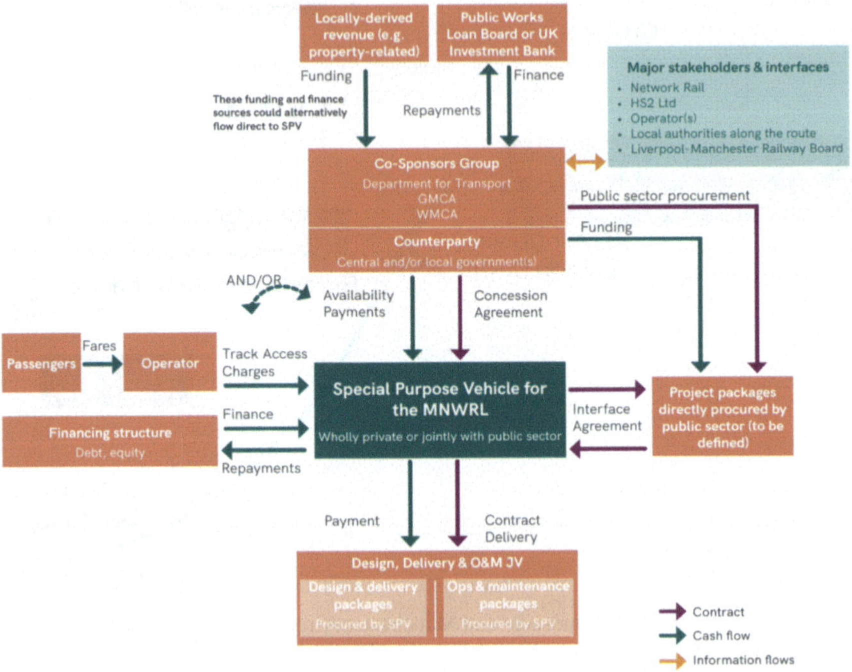

Fig. 10. Catalyzing economic growth through a Midlands-North West Rail Link, September 2024 [33]

Through these comprehensive modifications and strategic approaches, the proposed MNWRL is projected to achieve approximately 60–75% of the capital cost of the former HS2 Phase 2 scheme, while still maintaining a substantial proportion of the economic and transport benefits [33]. Further, this co-sponsorship approach will better explore the impacts of the project on economic development, decision-making, risk management and mitigation measures to support the project's objective. The approach is a bit complex as it needs to align multiple parties with the right vision and committed governance.

Analysis

The case of HS2 rescoping offers a detailed examination of the complexities involved in megaprojects and provides key insights into the economic, logistical, and political factors that have shaped its trajectory. Initially conceived as a transformative high-speed rail network, HS2 aimed to enhance connectivity between the North and South of England, reducing travel times, alleviating congestion, and contributing to economic rebalancing by linking major cities such as London, Birmingham, Manchester, and Leeds. The original vision for HS2 was ambitious, proposing a phased construction process. The first phase, linking London and Birmingham, commenced in 2017 and is projected to be completed by 2030. Phase two was intended to extend the route to Manchester and Leeds, with future extensions possibly reaching Scotland. However, over time, the scope of HS2 has been significantly reduced. The abandonment of the Leeds route in 2021, followed by the cancellation of the Manchester connection in 2023, reflects mounting financial and logistical challenges. These decisions have left the London-to-Birmingham segment and a spur to the West Coast Main Line (WCML) as the only surviving elements of the original plan.

The escalating costs played a critical role in this shift. The cost per mile of HS2 was reported to be several times higher than comparable European high-speed rail projects. The COVID-19 pandemic and rising inflation led to a dramatic increase in the project's total cost, with Phase 1 alone exceeding initial estimates for the entire HS2 scheme. The declining cost-benefit ratio further undermined the project's financial viability. This sharp decline in expected returns was a pivotal factor behind the decision to scale back the northern extensions of HS2. The project's rescoping not only highlights the inherent risks of large infrastructure undertakings but also raises questions about the long-term planning and forecasting methodologies used in major public projects.

The cancellation of the northern legs of HS2 signifies a strategic shift in the UK government's approach to transport infrastructure. In October 2023, Prime Minister Rishi Sunak introduced Network North, a new initiative designed to replace HS2's northern phases. Network North aims to redistribute the £36 billion saved from rescoping HS2 by investing in smaller, localized transport projects across the UK. This pivot reflects a move towards a more regionally focused transport strategy, prioritizing broad-based benefits over the concentrated impact of a single high-speed line. This decision has generated considerable debate. Proponents argue that Network North will deliver more widespread benefits, reaching more communities across the country. Critics, however, contend that the cancellation of the northern HS2 phases undermines the original objectives of the project, particularly its aim to improve connectivity and economic integration between the North and South of the UK.

In the aftermath of HS2's partial abandonment, regional leaders in Greater Manchester and the West Midlands have advocated for a new solution: the Midlands-North West Rail Link (MNWRL). This proposal, supported by a coalition of private and public sector stakeholders, seeks to connect Birmingham with Manchester, aligning with the Northern Powerhouse Rail initiative. The MNWRL is envisioned as a more cost-effective alternative to HS2 Phase 2, utilizing a blended funding strategy with public and private investment. The MNWRL represents a pragmatic response to the challenges faced by HS2. By adopting a phased approach, reducing design specifications (e.g., lowering maximum train speeds to 300 km/h), and leveraging existing infrastructure, the project aims to deliver a significant portion of the benefits promised by HS2 at a reduced cost (60-75% of HS2 Phase 2). This approach underscores the importance of adaptive planning in large infrastructure projects, where financial constraints necessitate flexible, scalable solutions.

Therefore, the case of HS2 illustrates the intricate balancing act required in delivering large-scale public infrastructure projects. The initial prospects as a transformative national project have been tempered by escalating costs, declining economic returns, and a volatile economic climate. While the rescoping of the project aims to redirect resources more broadly, it also leaves unresolved questions about long-term connectivity and regional development in the UK. The rescoping emphasis on cost efficiency and public-private collaboration, offers a potential paradigm for future infrastructure initiatives, signaling a transition towards more modular, agile, dynamic, and adaptive approaches to transportation planning that extend over decades to facilitate both planning and execution.

4 Results and Observations

The changing prospects of megaprojects across the globe such as HS2 (UK), HSR (USA) or NEOM Line (KSA) can offer valuable lessons for future endeavours like HSR in Australia, India, and the Middle East. It highlights the delicate balancing act required in delivering transformative projects. Unforeseen challenges, financial realities, and shifting political landscapes can all force adaptations [35]. It demonstrates the need for robust long-term planning to ensure projects remain viable amidst changing circumstances. The rescaling or rescoping may not be the story envisioned at its inception. However, the project's evolution underscores the increasing importance of adaptableness, partnership between public and private sectors, and an emphasis on cost efficiency in planning and delivering infrastructure projects. As we look to the future of megaprojects, these lessons will be critical in traversing the long road from vision to reality. The insights gained from the case study on HS2's evolution are valuable lessons for future mega projects. In conclusion, transformative projects necessitate (1) adaptive planning frameworks, (2) the flexibility to navigate unforeseen obstacles, (3) a robust delivery model, and (4) sustained stakeholder alignment throughout the project's trajectory.

4.1 Adaptive Planning Frameworks

It refers to an iterative approach to project management that enables adjustments in response to uncertainties and evolving contexts. Given the scale, complexity and extended durations of mega projects, rigid planning often results in inefficiencies or failure when unforeseen challenges arise. Adaptive planning entails predictive analysis, continuous monitoring, risk management, stakeholder engagement, and scenario-based forecasting to incorporate feedback and refine goals, timelines, or strategies as the project progresses. This feature can enhance resilience, optimize resource allocation, and improve the likelihood of achieving objectives in dynamic environments.

4.2 Flexibility to Navigate Unforeseen Obstacles

Given the complexity and extensive timelines, unforeseen challenges such as regulatory changes, environmental requirements, and socio-political distractions can significantly impact the project's success. To address these uncertainties, a project management approach emphasizing adaptive strategies, real-time data analysis, and contingency planning is essential. This enables project teams to reassess timelines, reallocate resources, and revise objectives when necessary. Flexibility also involves stakeholder engagement and dynamic risk management, which together ensure more resilient and responsive project execution. These methods mitigate risks and improve overall project sustainability.

4.3 Robust Delivery Model

Accountability and resilience are fundamental to achieving project objectives within the established timeline and budget. It is essential to identify a delivery model that aligns with the project environment and can integrate comprehensive risk management, and transparent governance. This approach serves to mitigate common failures such as cost overruns, scope creep and production delays. Historic indicate that public-private partnerships (PPPs) characterized by shared responsibilities and risks among stakeholders enhance project efficiency. Moreover, the integration of advanced project management principles like Agile practices, data analysis, machine learning and Building Information Modeling (BIM) can significantly improve decision-making and proactively identify potential failures before they materialize. Research shows that megaprojects with strong governance and well-defined lines of accountability are better positioned to avoid the systemic failures that have historically plagued large-scale endeavours.

4.4 Sustained Stakeholder Alignment

A strong alignment with upward and downward project stakeholders is crucial for the effective execution of megaprojects due to their inherent complexity and scale. Effective stakeholder management can influence project objectives, foster collaboration, mitigate risks associated with differing interests and enhance communication. Projects with well-defined stakeholder management strategies tend to achieve better alignment with regional and national needs, ultimately leading to more sustainable outcomes. For instance, aligning project objectives with stakeholder expectations encourages the co-creation of value, which can enhance the overall perception and acceptance of the project among local communities. Furthermore, involving stakeholders early in the planning and design stages can result in valuable insights, reducing the likelihood of conflicts that lead to delays later in the project lifecycle. Thus, incorporating a robust stakeholder alignment strategy is not just beneficial but essential for the success of megaprojects, enabling them to deliver on their intended objectives while fostering long-term positive impacts on the built environment.

5 Conclusion and Further Research

This qualitative research illuminates the critical challenges in transforming megaproject outcomes across various lifecycle stages. The findings emphasize the necessity for enhanced project conceptualization, incorporating comprehensive planning frameworks that extend beyond immediate stakeholders and conventional economic metrics. Risk assessment and cost-benefit analyses must adopt a more holistic approach, considering broader societal impacts and long-term implications.

Design flexibility emerges as a crucial factor influencing decision-making processes and change management strategies. The research highlights the significance of downstream consents and evolving environmental permits, which require meticulous consideration in transformational projects. Furthermore, the study underscores the importance of accurate factoring of economic indicators, including inflation rates and GDP, to ensure more reliable project forecasting.

Building on these insights, future research could explore the application of metaheuristic algorithms for enhanced planning and optimization. These advanced computational methods could potentially address the complex, multi-dimensional challenges inherent in megaprojects, offering innovative solutions for resource allocation, scheduling, and risk mitigation. Such algorithms could adapt to the dynamic nature of megaprojects, providing adaptive and robust optimization strategies across various project phases.

The findings suggest that by integrating these various elements - from improved conceptualization to advanced optimization techniques - future megaprojects could achieve greater success rates and more predictable outcomes. This research contributes to the growing body of knowledge on megaproject management, offering both theoretical insights and practical implications for project stakeholders and decision-makers.

References

1. Flyvbjerg B (2014) What you should know about megaprojects and why: an overview. In: PMI, From Academia: Summaries of New Research for the Reflective Practitioner
2. Garemo N, Matzinger S, Palter R (2015) Megaprojects: The good, the bad, and the better. Mckinsey & Company, Our Insights, New York
3. Flyvbjerg B (2019) Why megaprojects systemically fail - and what can be done about it. Oxford Answers. University of Oxford, Said Business School
4. Delphine, Patrick Witte & Tejo Spit, Megaprojects—An Anatomy of Perception, Local People's Perceptions of Megaprojects: The Case of Suramadu, Indonesia (2019)
5. Gil N, Beckman S (2009) Infrastructure meets business: building new bridges, mending old ones. Calif Manage Rev 51(2)
6. Contruent (2024) https://www.contruent.com/resources/blog/mega-projects-middle-east/. Accessed 27 Sept 2024
7. Buchholz K. The World's Megaprojects, published online. Statista. https://www.statista.com/chart/29653/megaprojects/. Accessed 28 Sept 2024
8. Statista. https://www.statista.com/chart/29653/megaprojects/. Accessed 28 Sept 2024
9. Neuroject. https://neuroject.com/construction-projects/. Accessed 28 Sept 2024
10. Rebim. https://rebim.io/largest-construction-projects-in-the-world/. Accessed 30 Sept 2024
11. The Urban Developer. https://www.theurbandeveloper.com/articles/eight-mega-projects-completing-2024. Accessed 30 Sept 2024
12. Statista. https://www.statista.com/statistics/1307599/big-infrastructure-construction-projects-worldwide-by-selected-countries/. Accessed 28 Sept 2024
13. Statista. https://www.statista.com/chart/3052/over-budget-construction-projects-in-comparison/. Accessed 28 Sept 2024
14. BBC. https://www.bbc.co.uk/news/articles/cyrr3mm8evyo. Accessed 26 Sept 2024
15. IEEE. https://spectrum.ieee.org/high-speed-rail-united-states. Accessed 26 Sept 2024
16. Statista. https://www.statista.com/chart/17070/evolution-of-the-total-budget-estimation-for-californias-bullet-train-project/. Accessed 26 Sept 2024

17. Network North. Transforming British Transport, Prime Minister's Foreword. p 5
18. HS2. https://www.hs2.org.uk/what-is-hs2/. Accessed 26 Sept 2024
19. Department of Transport (2017) UK Government, High Speed Two Phase Two Strategic Case, Moving Britain Ahead
20. BBC News. HS2: North West and Yorkshire routes confirmed. Accessed 15 Nov 2016
21. The Guardian. https://www.theguardian.com/uk-news/2021/nov/18/hs2-rail-leg-to-leeds-scrapped-grant-shapps-confirms. Accessed 18 Sept 2024
22. The Guardian. https://www.theguardian.com/uk-news/2022/jun/07/rail-bosses-outraged-as-hs2-golborne-link-quietly-scrapped. Accessed 15 Sept 2024
23. BBC News. Rishi Sunak speech, PM scraps HS2 link, saying 'facts have changed'. Accessed 4 Oct 2023
24. Independent. https://www.independent.co.uk/news/uk/hs2-prime-minister-manchester-west-midlands-crewe-b2431486.html. Accessed 20 Sept 2024
25. Network North. Transforming British Transport, Prime Minister's Foreword. p 14
26. PwC. High Speed Rail International Benchmarking Study. Accessed Nov 2016
27. Network North. Transforming British Transport, Prime Minister's Foreword. p 13
28. Network North. Transforming British Transport, Prime Minister's Foreword. p 15
29. Network North. Transforming British Transport, Prime Minister's Foreword. p 26
30. New Civil Engineer. https://www.newcivilengineer.com/latest/new-line-between-handsacre-and-manchester-airport-proposed-to-replace-hs2-phase-2-21-03-2024/. Accessed 22 Sept 2024
31. HS2. https://www.hs2.org.uk/in-your-area/assistance-for-property-owners/project-rescoping-phase-two-cancellation/. Accessed 21 Sept 2024
32. New Civil Engineer. https://www.newcivilengineer.com/latest/mayors-and-civil-engineers-call-for-conventional-rail-link-in-place-of-hs2-phase-2-16-09-2024/. Accessed 22 Sept 2024
33. Opportunity through connectivity, Catalyzing economic growth through a Midlands-North West Rail Link. Accessed Sept 2024
34. Dumitriu S, Hopkinson B (2023) Britain's infrastructure is too expensive, August 2023
35. Omega Centre (2012) Mega Projects Executive Summary. In: Lessons for decision-makers: an analysis of selected international large-scale transport infrastructure projects

Mega Science and Re-Nationalization in the Field of Semiconductors. An Evidence-Based Exploration of Decoupling Triggered by the China Initiative

Giulio Marini[1](✉) and Sara Lombardi[2]

[1] University of Catania, Catania, Italy
giulio.marini@unict.it
[2] University of Florence, Florence, Italy

Abstract. Few technologies have attracted as much concern recently as semiconductors, known for being among the most expensive endeavors in science and technology. There are at least two concurrent reasons for this heightened attention: first, the United States is striving to defend its supremacy against a rising China through policies such as the China Initiative and the CHIPS Act. Second, Taiwan, the world leader in semiconductor frontier and production, is a highly politically contested region. By analyzing collaborations across Mainland China, Taiwan, and the US, this study explores the impact of research megaprojects on scientific publications. We specifically investigate the extent to which decoupling is occurring, identifying which players—among the locations of authors, funding agencies, and major global private companies—are more likely to produce leading research in the field, particularly before and after the implementation of relevant US policies. The results indicate that, while we cannot establish causal effects, decoupling is indeed taking place, with the US side being comparatively the most detrimentally affected by its own policies.

Keywords: Mega Science · megaprojects · semiconductors · decoupling · re-nationalization

1 Introduction

Mega Science (MS) is a conventional term to refer to the growing-ray that science needs to keep in order to produce notable discoveries (Baker & Powell, 2024). An important factor in the post-WWII scientific regime is that of the interplay of technology and science itself, which paves the way to both fruitful positive externalities and constraints. It is also a matter of organizing science, with shooting up numbers in personnel involved and costs of infrastructures. It also defines pertinency to megaprojects (Flyvbjerg, 2014). In other words, science is perhaps the blueprint of why megaprojects are important to be studied, as MS strictly implies referring to megaprojects, at least if conceived as a series of intertwined projects. Science might be blue sky or applied, or even "Pasteur

E. Favari et al. (Eds.): MeRIT 2024, LNCE 600, pp. 228–243, 2025.
https://doi.org/10.1007/978-3-031-92797-3_15

quadrant" (both highly innovative and highly with important applications). Science is public per se, as the competitive and highly stressful game of publishing consists in being the first to solve a problem, being the first in advancing in something, being the frontrunner of a new stream of promising research, establishing as a watershed in some field, etc. Whatever the tracer of such success, the game is not just that of getting something – a *private* good. It is instead that of showing and disclosing – a *public* good. In this sense, to publish science is both highly abstract and potentially "useless", but at the same time, at least to some regards, it is a "competitive advantage dissipator": once something is discovered, it is inevitably understood by others, rivals included. As far as the rivals are other colleagues working on the same or similar issues, the game remains within the science sector. Nowadays, rivalry at country level has become that of China catching-up the US in terms of frontier, claiming to be the main superpower along with the US itself. In fact, whenever the implications go beyond individuals' intellectual glory, sovereign countries and companies may re-emerge as players with vested interests. These latter players also fund such MS although countries represent the main source of funding. Whatever may happen within science-makers, those who end to put one's name as authors in scientific publications may enter into conflict with those having aforementioned vested interests – either they are companies seeking advantage against other companies or countries seeking to arrive before others to some results of note.

Based on this premise, this paper aims at analyzing whether, in recent years, policies to decouple China from the US produced any actual change. We control so via scientific collaborations across the main two superpowers in science, Mainland China and the US, and respective quality of research. We include the relatively minor region of Taiwan to investigate further this conundrum, as the region rose to prominence by means of a national strategy whereby universities had a pivotal role in (Hsu & Yuan, 2013).

The paper is structured in the following way. The next section describes the issue of collaborations in science, building some more insight into a key player – funding agencies. It then depicts how MS is affected by globalization and its apparent possible opposite – decoupling or re-nationalization, as well as the main policies that have developed the field, appreciating the notable role of national policies. The section concludes with more contextual information about the three regions we discuss: US, Mainland China, and Taiwan. The rest of the paper deals with empirical analyses, presenting preliminary findings about the effect that US policies, particularly the China Initiative one, might have produced so long on quality of science, bearing in mind that such outcomes of broad policies go beyond the remit and the capacity of a single study.

2 Theoretical Background

2.1 Competition, Collaboration, and Human Factor

The intrinsic dilemma of collaboration in science and technology has been already dissected in literature. Competition and collaboration may coexist within the same pair of rival companies, when working on, say, patents. If this applies to patents and companies, which are, by definition, conducive and enactor of business (Yoon et al., 2021), a fortiori universities and public (and not public) research institute may work together despite being affiliated in countries that are rival on a general basis, such as Taiwan vs. South Korea, or China vs. US. This argument is reinforced by the fact that Yoon and colleagues research is realized on semiconductor industry itself. However, the fact that patents are in principle more practice and business oriented should not let fall into the bias that such metric alone could represent industry prowess. Filippin (2021) finds such surprising evidence crunching data along the Moore's Law. Moreover, spillover effects in patenting in the field has been found to have non-random networking latency, with some national pattern within it, when it comes to talk about companies (Wang & Huang, 2017). Marini and Mouritzen (2024) pose the issue under game theory, speculating that decoupling may have a sense if the decoupler has realistic expectation to lose less than the possible partner will.

It is widely acknowledged that development of science has implied, inter alia, also a more frequent practice of international collaborations, typically measured by location of affiliations of authors (Aksnes & Sivertsen, 2023; Leydesdorff & Wagner, 2008). This established stream of research, mostly developed around science and technology, bibliometric analyses, sociology of science, and higher education, has unveiled an apparently unlimited densification of international collaborations. Putting in simple words, the Michael Polanyi's idea of a "Republic of Sciences" (RoS) and the idea of "Invisible Colleges" (Wagner, 2008) underlie that any scientist in the world, regardless of any characteristics, nationality of scientist herself and nation where she is based, is able not only to read publications by others and propose hers, but also to work together. Possible conflicts between two countries are irrelevant for the sake of science advancement, under this argument. This assumption is coherent with Mertonian norms of communism, universalism, disinterestedness, and organized skepticism. None today would question rivalry between scientists based in France and Germany as it could have been once upon a time. However, total freedom of collaboration may result questionable in some few cases that might be relegated to exceptions. For instance, to work on atomic energy for civil applications with colleagues based in, say, Iran or North Korea may result as inappropriate, dangerous, and in any case hyper-regulated and controlled by international bodies. Hyper-regulation itself may discourage collaborations, as one may prefer to simplify her job.

One may then conclude that international collaborations are conducive of the utmost and finest research. In fact, internationality per se is not associated with higher degrees or likelihood of creativity and novel findings (Wagner et al., 2019). Also bearing this backdrop in mind, which deals with counter-intuitive but also marginal gains, international collaborations in MS are, statistically if not normatively, the norm.

2.2 The Role of Funding Agencies in MS

There is no science without a huge amount of money. The US research capacity is aptly identified by the connection between non-University funders and therein regulations to systematically involve faculties – the researcher dimension of the academic staff in, mostly but not exclusively, research-intensive universities (Baker & Powell, 2024). This connection is arguably a dynamic we may interpret as an element of a function. In fact, research outputs (publications) are a function of the intrinsic desire by the RoS to publish the most interesting, cutting-edge results from one side, and the desire expressed by a

State via some new public management (NPM) agencies. These agencies are research Funding Agencies (FA) shaping what to investigate under the incentive of giving money and other enablers to do research, such as access to extremely expensive facilities. In some circumstances, FAs develop programs to incentivize periods abroad purposedly to increase collaborations, mostly devised to acquire critical knowledge from the outer world, for thus coming back and cashing the knowledge surplus (Quan et al., 2019). While the potential for collaboration between RoS and FAs is promising, problems arise if FAs prefer that RoS limit the development of scientific knowledge, impeding scientific production under certain circumstances. This could lead to the exclusion of researchers located in undesirable regions, or having passport issued by suspicious or rival country. These possibilities meet what Chine Initiative has represented despite actual consequences of such policies are uncertain.

The idea that Funding Agencies are capable of shaping science is consolidated in the literature (Veletanlić & Sá, 2020), although the role of companies as funders per se is less consolidated and mostly included in University-Industry stream of research. Some supranational FAs are able to favour expansion of international collaborations – the European Commission being probably the most notable point in case. Much less is known in relation to FA preventing, via nested regulations, collaborations with specific locations under the principle of exclusiveness and strategic reasons one may synthesize with the expression *Raison d'État*. This factor is surprisingly unexplored in literature. In MS, the idea of problems arising from the "human factor" is widely known, but it represents in the first place the (un-)desired consequence of "boundary deactivation" (Rüland, 2023). Re-nationalise science, if it really does occur, may represent a loss at aggregated levels following this argument, as some coalition-building across countries might be discouraged. In these most recent years, the case in point is that of the US and China.

2.3 The Current (Post-?) Global World

The idea that science is pivotal for national interests is somehow self-evident. The fact that free international collaboration may generate Pareto optimal outcomes is as well conceivably true. The emerging problem in recent years has been the decoupling thesis whereby countries may perform actively to prevent others from overcoming them in scientific development. Decoupling is commonly understood literally as avoiding being dependent on others via developing one's, say, technology. Interestingly, the US-China dynamic, the key one of the current century, has already showed both increased and decreasing decoupling, giving no clear pattern whether globalization as publicized in late 90s and early 2000s is over or not (on a marginal note, it is interesting to note that Francis Fukuyama's forefront to *Invisible Colleges* by Caroline Wagner in 2008 was not envisaging at all such type of rivalry among the list of "bad globalizations" he discussed). For instance, Han et al. (2024) "observe signs of decreasing decoupling despite restrictions on travel and strains on supply chains" for the post-tension years (2020 year onward). Under this global context, China is navigating not only its own growth, but also the global consequences of such growth (Marginson, 2022). Global West, on the other hand, is still prominent and represents the frontier, for instance in health in relation to Covid-19 (Marini, 2024), although reasons to believe that China is the frontier in some domains are possible, if not already a reality. Under the lenses of global competition, evidence support that processing electric power capability will be a critical limiting factor on the ability of companies and countries to develop advanced AI systems – AI requiring computational power which progresses speedier than Moore's Law itself (Miller, 2024).

In this case, the problem of competition is operationalized as output (publications) being litmus test of FA eventual capacity to prevent RoS to collaborate with potentially risky partners, where risk is that of giving to rivals critical advantages that would later rather than sooner acquire (i.e., the "leaking knowledge" argument that, applied to science, equals to get as soon as possible knowledge as a driver to more quickly advance also in applied research).

2.4 The Scientific Race in The Field of Nanotechnology

It is relevant to build awareness of how nanotechnology developed into megaprojects. Some scientometric analyses show that semiconductors laser publications had been dominated by US affiliated institutions up to the 70 s, for later witnessing the surge in this strategic sector of the emerging superpower of the 80 s and 90 s – Japan, a country which was also synonymous of techno-nationalism in those years. Up to roughly 2000-year, American corporations were at the forefront of science frontier in the sector, AT&T for instance outscoring also the whole University of California (Tsay et al., 2000). This stage also paved the way to a sort of MS strongly dominated by international and industry-university partnerships, which were visible and traceable via publications (Shimizu, 2011), and geographical spin-offs pushed by mobility of researchers (Wai, 2023). Until end of XX century though, around half of semiconductors related publications were signed by US-affiliated researchers, with China accounting for only 1.28%, and half of these publications were single authored (Tsay et al., 2000). Global, meaning international development, and "techno-nationalism", meaning countries developing their own roads, alternate each-other routinely, having some countries entered a re-national one almost a decade ago (Park, 2023). As it would be intuitive, to come back to the national dimension, at least in "sensitive industries", highly covariates with re-launching nation-wide, rather than international, projects – that's the case for instance of Germany and digital infrastructure (Möllers, 2021). Industry-university partnerships have been always highly regulated, bringing from early stages to consortia and other forms to aggregates efforts on an ever-larger scale, and involving universities (Lécuyer, 2022; Mody, 2017). The CHIPS Act, it is reported (Chapman, 2024; Peters, 2022), is mostly a matter of incentives for national security pivoted in R&D, in fact. CHIPS Act is also intentionally rooted in the idea of preventing China to dominate the strategic sector of semiconductor which is, economically, conducive of large-scale applications. However, the role of scientific research is not directly traceable from these strategic political actions, as at least agencies involved are multiple. The most relevant US funding agencies in the research field of interest (nanotechnology) is the Department of Energy, and the other main one being the National Science Foundation. It is nevertheless manifest that bipartisan "Science Act" passed in 2021–2022 is supposed to be a turning point for the US to secure enduring leadership in this specific strategic scientific domain. The Chinese side of this race might be summed up by the evidence of a high deficit of self-reliance in early 2000s up to a fair betterment in terms of degree of US dominance in semiconductors, if measured by patents (Han et al., 2024).

2.5 Evidence of US – Taiwan – Mainland China Collaborations in Nanotechnology

It is known in literature that the US and China have different approaches when it comes to refer to scientific collaborations: more international and diversified the former, more domestic or relying on mostly OECD countries the latter (Jang et al., 2022). Some first hints of such a relative decline of commitment of international collaborations already occurred (Shu et al., 2022), but such study is built on aggregated figures and needing updates, conceptual unpacking, more analytical breakdown. A broad decoupling thesis has recently dominated wide public debate, including the issue of the value chain of semiconductors (Grimes & Du, 2022). In this particular sector, China is still backward though highly committed in keeping collaborations and reducing the delay in technology. Research collaboration is arguably one of the main gateways, if not the most important, to advance and to fill such a gap. Quests of "sovereignty" pertain also Europe, despite apparently awareness of neo-national competition appears to be laggard relative to the US or other main Asian Countries (Johnston & Huggins, 2023).

Some more insight is required in relation to the region of Taiwan (self-declared "Republic of China, ROC" but nowadays rejected widely from international debate under the doctrine "there is only one China"). From the US point of view, Taiwan is historically an ally and is now representing a threat of almost monopsony when it comes to talk about semiconductors, which may turn to be a critical chock point in case of supply disruptions. From the Chinese point of view, "The Strait" issue and respective diplomacy have for decades shaped political interests. It is also a topic of utmost importance in political discourse. The issue is relevant in science as well, as Taiwan presumably has an interest in keeping monopoly on top-tier technology in the field of microprocessors, whereas certainly Mainland China has only recently rose in such industry to almost put itself at the par to global leaders. In semiconductor production, Taiwan has surged to global leadership in the field via very consistent R&D investments for which Science Parks are known to be a key super-structure to thrive University-Industry collaborations (Hu, 2011). As a matter of fact, an umbrella organization like the National Applied Research Laboratories in Taiwan publishes around the double of research than notorious Taiwan Semiconductor Manufacturing Company Limited (TSMC, a spin-off itself of Hsinchu Science Park) alone does, representing a bridge-institution tightly linked to national interests. Industrial Technology Research Institute of Taiwan and the National Synchrotron Radiation Research Center are both even more prolific in research. These few figures let believe that in comparison to companies, these blended institutions are more relevant when assessing science production per se.

As already discussed, much of development in science is due not only to funding, but also to either/both collaboration or/and brain drain. We focus on collaborations meant as co-authored publications, bearing in mind the relevant distinction/connection between a highly research-intensive industry from one side, and publishing research on the other side – for a thorough depiction of the successful Taiwanese model and the role of academic R&D (see (Chen & Jan, 2005)). Tensions due to Strait diplomacy establish a dilemma in terms of either collaborating (which has win-win features) or not (e.g., to prevent rivals to catch-up sooner rather than later).

3 Data and Methodology

We propose a bibliometric analysis of collaborations across the three entities (Mainland China, Taiwan, and the US; CN, TW, US respectively) in the field of nanotechnology – a field known for encompassing some the most strategic development in science for respective applications, including semiconductors. Data are downloaded for Clarivate Web of Science, accounting for location of authors (CN, TW, US), location of funding agency (CN, TW, CN; the main ones for each region – see further), international co-authorships or domestic publication (the latter identifies possible co-authorships, but only within same country of affiliation of authors), industry collaborations (publications co-authored by any of the main 26 companies[1] in the field of semiconductors), and a series of conventional bibliometric measures. Despite international networks are also a function of researchers' mobility (Finocchi et al., 2023), we solely look at performed networks, appreciating that countries incentivizing returnees may roll over also this side advantage in terms of extra-national collaboration (Chu et al., 2023; Hsu et al., 2023; Marini & Yang, 2021).

The idea of inserting also companies deserves a justification. This choice meets freshest evidence whereby industry-universities links have become increasingly key in the semiconductor networks, especially when facing the issue of Government subsidies (Chang, 2024). At the best of our knowledge, industry-university literature is mostly focused on patents, whereas here we point to publications – the epitome of scientific output and possibly a less attracting effort for corporations that should pursue an eventual effort to translate such research into something entering within their conventional remits. Moreover, subsidies differ from (research) funding agencies. One of the few exceptions on this dearth of knowledge highlights that corporations' scientific publications provide feedforward beneficial effects on same corporations' patents, enacting a virtuous circle especially in early stages of developments (Chen et al., 2024). This evidence also implies that any effect of involvement in publishing by corporations has, as expectable, long-term, rather than short-term, impacts. More specifically "corporations operating in the semiconductor industry tend to generate a greater quantity of patents with improved quality when they enhance the utilization of joint scientific research outputs as references in their patent citations" (Chen et al., 2024: 11), which also means that connections with other actors, more likely universities, are more conducive to profitability. However, ex-post considerations point to the necessity to better understand science parks roles in accounting affiliation of scientific publications.

The number of authors per scientific output is a confounder deserving specific argument. The number of co-authoring people is at the basis of the concept of MS (Kuperman & Sokol, 2024). Its relevance is also reinforced by China hosting more often the

[1] Samsung Electronics, Intel, TSMC, SK Hynix, Micron, Qualcomm, Broadcom, Kioxia, Texas Instruments, Analog Devices, Microchip, NXP, MediaTek, Infineon, Bosch, STMicroelectronics, Sony, ARM, AMD, Nvidia, ON Semiconductor, UMC, Apple, IBM, Mitsubishi Electric, Tower Semiconductor, Xilinx, SMIC, Nordic Semiconductor, Semi-Conductor Laboratory.

"dominant" role in multinational collaborations, despite having a declining share if looking at the time series – papers having at least one author with multiple affiliations in more countries (Liu & Huang, 2022). Conversely, "contributing" collaboration from the US is conducive of more quality, albeit with hints of being less frequent. We use this variable to understand whether engagement with international colleagues might be not justified by the number of authors. If the number of authors is high and increasing, presumably likelihood of internationality ought to proportionally increase. On the other hand, an increase of co-authoring researchers without proportional involvement of colleagues from abroad is likely to be a sign of "(re-)nationalization" of science.

Methodology is moving first from some time series and description of the field by locations of authors and location of funding agencies. Thus, we seek to predict quality of publications captured by Percentile in Subject Area (PSA), which is only feebly time sensitive. Last step of analysis is that of approximating some quasi-experimental research design to test whether post-tension period, identifiable as "post China Initiative" (2020 included onwards), has changed patterns of quality of publications, under the reasonable assumption that US funding agencies roll over effects of national policies related to hypothesized re-nationalizations.

4 Results

Prima facie analysis show that, in nanotechnology, Taiwan is, despite its global success in the field, inevitably smaller than US and Mainland China (see Figs. 1 and 2). As for Taiwanese production, Table 1 shows that authors located there used to collaborate frequently with authors affiliated in the US, but such growth ceased, if not declined, in most recent years. Although these are not definitive signs, there is room to believe

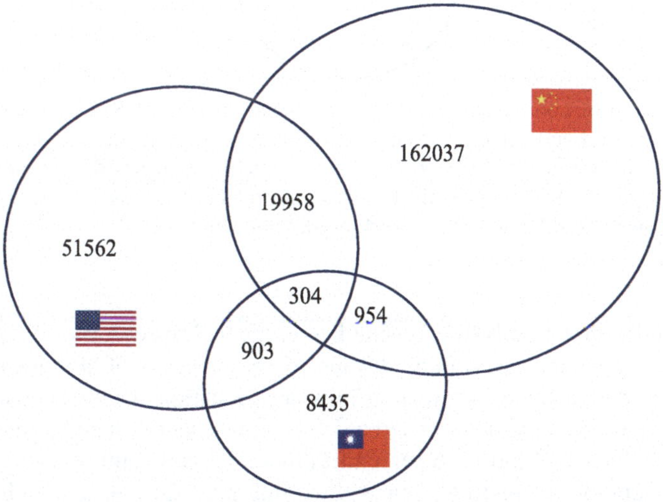

Fig. 1. Distribution of collaborations by location of authors (affiliation). Number of publications. Each subset is suitable for further international co-authorships, here overlooked. Publications in nanotechnology 2008–2023, Web of Science.

that there might be some decoupling between the US and Taiwan. On the other hand, Taiwanese researchers seldom co-authored with colleagues in the Mainland in first years under observations, for quickly increasing such inter-dependence. In most recent years, Mainland China surges for Taiwanese authors to be the most frequent collaborating partner. Table 1 also reports the times series by locations of either authors or funding agencies between Mainland China and US. 2020 year is the turning point if considering location of authors, and 2019 year if location is that of funding agency. It means that the increasing partnership in science between US and Mainland China decreased in last years in the field. This evidence reinforces the idea that funding agencies are first-factor in driving changes, namely decoupling in this case. That of declining collaborations between Mailand China and US is a preoccupation Tang (2024) spotted more widely in science as figures have become visible in early 2024. In this regard, Tang adds a normative stance on the detrimental consequences of such overall evidence.

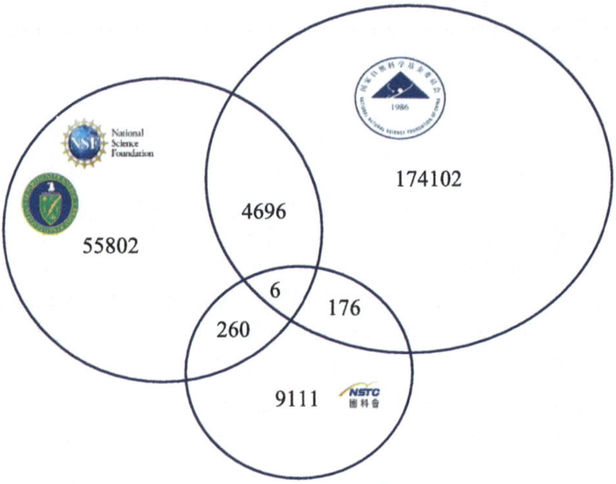

Fig. 2. Distribution of collaborations by funding agency (for the US: Department of Energy and National Science Foundation; for Mainland China National Natural Science Foundation of China; for Taiwan: National Science Technology Council). Intersects denote publications co-funded by multiple agencies irrespectively to location of authors. Each subset is suitable for further international co-authorships here overlooked. Publications in nanotechnology 2008–2023, Web of Science.

We proceed with the analysis choosing Percentile in Subject Area (PSA) as a dependent variable to predict success in publishing by the interplay of aforementioned variables. OLS reported in Table 2 shows that Taiwanese location of authors do not represent a statistically significant predictor, whereas US is more effective than Mainland China in secure better influence (almost 3.6 percentiles more; Chinese author locations increase 2.7 PSA). In relation to funding agencies, we interpret Taiwanese deficit as a scale-derived effect – no region can par the US or China, unless perhaps grouping Europe as a whole. US and Mainland Chinese funding agencies instead are of similar predictive capacity. As expected, domestic publications have a negative impact, and the number

Table 1. Times series of publications in nanotechnology between US-CN, excluding Taiwan by: (a) location of authors; and (b) funding agency location. Columns (c) and (d) provide co-authorships between authors located in Taiwan and US (c) and Taiwan and Mainland China (d). 2008–2023, Web of Science.

Year	(a) - CN-US (Authors)	(b) - CN-US (Funding Agency)	(c) – TW-US (Authors)	(d) – TW-CN (Authors)
2008	90	27	12	6
2009	238	63	50	13
2010	346	88	68	20
2011	433	121	62	21
2012	591	187	106	31
2013	744	197	81	48
2014	893	256	79	52
2015	1223	307	82	53
2016	1522	422	73	63
2017	1881	490	77	89
2018	2251	564	77	108
2019	2453	554	96	151
2020	2139	466	95	146
2021	1881	411	101	144
2022	1569	296	78	138
2023	1393	221	58	154

of co-authors is positive. Interestingly, the presence of any of the most important global private companies dealing with semiconductors lessens the likelihood of having relevant publications in nanotechnology.

Table 2. OLS Prediction of PSA on all nanotechnology publications between 2008 and 2024 by any author affiliated in either US, Taiwan, or Mainland China and those funded by any of NSF, DOE, NSTC, and NSFC*.

Variable	Label	coef.	S.E.
loc_TW	Publication authored by at least one person affiliated in Taiwan	0.334	(0.90)
loc_CN	Publication authored by at least one person affiliated in Mainland China	2.685***	(0.47)
loc_US	Publication authored by at least one person affiliated in the US	3.555***	(0.29)

(*continued*)

Table 2. *(continued)*

Variable	Label	coef.	S.E.
NSTC	Publication funded by National Natural Science and Technology Council (CN)	−7.742***	(0.99)
NSFC	Publication funded by National Science Council (TW)	2.292***	(0.54)
NSF	Publication funded by National Science Foundation (US)	1.330***	(0.31)
DOE	Publication funded by Department of Energy (US)	2.150***	(0.29)
anycorp	Publication authored by people affiliated in any of most relevant companies	−6.736***	(1.21)
dom	Domestic publication (location of authors)	−2.792***	(0.17)
nocoau	Number of co-authoring people in given publication	1.595***	(0.02)
Year	Year of publication	52.666***	(0.77)
_cons		0.334	(0.90)
N		244,153	
R2		0.1575	

**Publications for 2024 year are not complete*

Considering the recent issues discussed in this paper about making collaborations more difficult in sensitive fields like semiconductor, we test the extent to which two different periods provide different results. Table 3 splits between before and after the moment China Initiative is believed to have entered in its effect period, also in coherence with Table 1.

Table 3 reveals to be heuristically useful. The most notable differences in coefficients between pre and post (or during) China Initiative is that US location of author decreased, losing almost 2 percentiles. Also, location of authors being Mainland China results in having some lower likelihood to have better research, but in this case the loss is inferior. This might be read as decoupling damaging all players (i.e., both the US and Mainland China), but causing more damage on the American side of the Pacific Ocean rather than on the Asian one. Even more interestingly, funding agencies as predictors change with clearer sign. If in pre-China Initiative period (2008–2019 included) US Department of Energy and National Science Foundation were predicting on average respectively 4.2 and 2.3 higher PSA, main Chinese counterpart in the field was securing only 1.1 increase of PSA. In "China Initiative" period, the current one from 2020 year included onwards funding agencies' capability to increase PSA is reversed, with the Chinese one (NSFC) resulting in performing better than the other two American ones. This stage of analysis is not an analysis suitable to claim any average treatment effect on the treated as per conventional quasi-experimental research design. In fact, it could be that without China Initiative, provided such wide program is having any effect on the data we are treating, the divide we find here might have been even wider.

For seeking a tentative answer to our question about whether in the field of nanotechnology China and US (and Taiwan as an in-between player in science) are decoupling,

Table 3. Before (2008–2019) and after (2020–2024) OSL predictions of PSA in nanotechnology field. Labels of variables as per Table 2*.

	Before		After	
	coeff.	S.E.	Coeff.	S.E.
loc_TW	−3.130**	(1.15)	4.069**	(1.42)
loc_CN	2.806***	(0.54)	2.695**	(0.91)
loc_US	4.943***	(0.36)	1.887***	(0.47)
NSTC	−4.904***	(1.24)	−7.258***	(1.67)
NSFC	1.304*	(0.61)	4.467***	(1.05)
NSF	2.060***	(0.35)	−0.625	(0.60)
DOE	3.592***	(0.33)	−0.649	(0.55)
anycorp_du ~ y	−7.575***	(1.31)	−5.328*	(2.67)
dom	−2.122***	(0.22)	−3.528***	(0.27)
nocoau	1.810***	(0.03)	1.382***	(0.03)
year	YES		YES	
N	135,896		108,257	
R2	0.0614		0.2201	

Publications for 2024 year are not complete

and, if yes, with which consequences, in Table 4 we provide a difference-in-difference test that uses any of the two US funding agencies here taken under analysis as a treatment variable, and keeping pre-2019 and 2020 year onwards as the most reasonable moment in time any change triggered by US policy-makers has occurred. In this case, treatment variable reads like US funding agencies might have worked under different regulations during Trump Administration years, producing visible effect from 2020 onwards. The hypothesis is that such changes pivoted in restrictions cause effects possibly traceable as an outcome in relation to PSA – namely, lower quality. Table 4 shows a large negative effect for DOE or NSF funded publications after 2019, which is almost one decile (10 percentiles) less good research. This means that once we account for some of the most relevant confounding variables, US funded research appears to lose capacity to be the best enabler of science. As per previous before-after analysis, this test is as well subject to a series of considerations preventing to maintain that we are finding a causal effect. For making an example, these data do not encompass any effect related to mobility of talents, which is presumably not less important than free collaborations. We also overlook hidden science, that that is more likely to fall under confidentiality for defense purposes, and we may allow even that such changes in regulations might need longer time to enter into full effect.

Table 4. Difference in Difference test for any of the two most important US funding agencies in the field of nanotechnology (DOE and NSF) by pre- and post- "China Initiative" periods in predicting PSA of publications*.

	Coeff.	S.E.
Treat (any between DOE and NSF funded publication)	4.923***	(0.49)
Period (2008–2019; 2020–2024)	−61.174***	(0.74)
Treat * Period	−9.303***	(0.30)
loc_TW	0.180	(0.90)
loc_CN	2.749***	(0.47)
loc_US	3.696***	(0.30)
NSTC	−6.899***	(1.03)
NSFC	2.255***	(0.60)
anycorp_du ~ y	−6.907***	(1.21)
dom	−2.795***	(0.17)
nocoau	1.594***	(0.02)
year	YES	
N	244,153	
R2	0.1606	

**Publications for 2024 year are not complete*

5 Conclusions

Our analysis timely sheds light on a hot phenomenon which has recently lit worried attentions, also for such attentions being interwoven with (1) huge investments dealing with megaprojects, and (2) extremely complex political issues. Many speculations have been related to supply chain, and next top-notch steps in semiconductors for securing critical applications. However, less is known when it comes to delve into the MS side of it, especially when science is not understood strictly in term of patents. We find some tangible effects of de-coupling or re-nationalism, which is likely to be a reaction different forms of "realism" (Luo, 2022) – a dynamic we may agree is likely to have below-zero effects relative to full openness, albeit requiring more investigations.

We believe that under CHIPS Act, but also under China Initiative, it is urgent to ponder empirically the outcomes. In relation to this wide research program, Aghion et al. (Aghion et al., 2023) provides demonstration of intended consequences of China Initiative related to Chinese scholars previously collaborating with American ones, put in comparison to Chinese scholars collaborating with European ones. They show that Chine Initiative is working as per US policy-makers desiderata. Our study explores a different angle, deploying a different research design in the first place. We arrive to some preliminary and provisional conclusions that (Mainland) China is not suffering this new stage of competition presumably implemented by funding agencies' new regulations. It seems it is sustaining effectively research in the field, instead, playing an increasing leading role in funding research in nanotechnology. Further investigations may question more tightly whether US intended restrictions might even accelerate Chinese catch-up and frontier-making status in the field. Within this crisis of balance among the two main superpowers (US and China), Taiwan is probably the most delicate in-between region, not only geographically and politically, but also for the semiconductor sector representing a further reason of concern. We conclude to this latter point that we recommend distinguishing scientific publications from manufacturing and economics, and also distinguishing whole science from patents – two aspects there is larger corpus of research about. Presumably, most of the research in Taiwan in the field is not explainable from the National Science and Technology Council funding agency only, but also from the interplay with Science Parks and other institutes linked to universities. Also, mobility of brains is likely to be a relevant factor, which is here not investigated, representing another relevant limitation.

References

1. Aghion P, Antonin C, Paluskiewicz L, Stromberg D, Sun X, Wargon R, Westin K (2023) Does Chinese research hinge on US coauthors? Evidence from the China Initiative? In: CEP Discussion Papers dp1936. Centre for Economic Performance, LSE. https://EconPapers.repec.org/RePEc:cep:cepdps:dp1936
2. Aksnes DW, Sivertsen G (2023) Global trends in international research collaboration, 1980–2021. J Data Inform Sci 8(2):26–42. https://doi.org/10.2478/jdis-2023-0015
3. Baker DP, Powell JJW (2024) Global mega-science: Universities, research collaborations, knowledge production. Stanford University Press, Stanford, CA
4. Chang SH (2024) Key technology and development of industry–academia collaboration patents derived from the government subsidy program. Sci Public Policy 51:748–759. https://doi.org/10.1093/scipol/scae015
5. Chapman B (2024) Recent US Government and foreign national government and commercial literature on semiconductors. J Bus Financ Librariansh 94-124
6. Chen JH, Jan TS (2005) A variety-increasing view of the development of the semiconductor industry in Taiwan. Technol Forecast Soc Chang 72(7):850–865
7. Chen X, Mao J, Ma Y, Li G (2024) The knowledge linkage between science and technology influences corporate technological innovation: evidence from scientific publications and patents. Technol Forecast Soc Chang 198. https://doi.org/10.1016/j.techfore.2023.122985
8. Chu JY, Chiu SY, Syu YS (2023) Returnee and local doctorates on research productivity: Doctoral training, international mobility, and research collaboration. High Educ 87(4):963–989. https://doi.org/10.1007/s10734-023-01046-4
9. Filippin F (2021) Do main paths reflect technological trajectories? Applying main path analysis to the semiconductor manufacturing industry. Scientometrics 126:6443–6477. https://doi.org/10.1007/s11192-021-04023-9
10. Finocchi I, Ribichini A, Schaerf M (2023) An analysis of international mobility and research productivity in computer science. Scientometrics 128:6147–6175. https://doi.org/10.1007/s11192-023-04837-9
11. Flyvbjerg B (2014) What you should know about megaprojects and why: an overview. Proj Manag J 45(2):6–19
12. Grimes S, Du D (2022) China's emerging role in the global semiconductor value chain. Telecommun Policy 46(2):101959

13. Han P, Jiang W, Mei D (2024) Mapping U.S.–China technology decoupling: policies, innovation, and firm performance. Manag Sci. https://doi.org/10.1287/mnsc.2022.02057
14. Hsu DW, Yuan BJ (2013) Knowledge creation and diffusion of Taiwan's universities: knowledge trajectory from patent data. Technol Soc 35(3):172–181
15. Hsu GC, Lin WJ, Sun SJ (2023) Temporal trends in academic performance and career duration of principal investigators in ecology and evolutionary biology in Taiwan. Scientometrics 128:3437–3451. https://doi.org/10.1007/s11192-023-04710-9
16. Hu MC (2011) Evolution of knowledge creation and diffusion: the revisit of Taiwan's Hsinchu Science Park. Scientometrics 88:949–977. https://doi.org/10.1007/s11192-011-0427-5
17. Jang B, Choung JY, Kang I (2022) Knowledge production patterns of China and the US: Quantum technology. Scientometrics 127:5691–5719. https://doi.org/10.1007/s11192-022-04478-4
18. Johnston A, Huggins R (2023) Europe's semiconductor industry at a crossroads: industrial policy and regional clusters. Eur Urb Reg Stud 30(3):207–213
19. Kuperman VY, Sokol GH (2024) On the causes and ramifications of multi-authorship in science. Scientometrics 129:2205–2225. https://doi.org/10.1007/s11192-024-04963-y
20. Lécuyer C (2022) Driving semiconductor innovation: Moore's law at Fairchild and Intel. Enterprise Soc 23(1):133–163
21. Leydesdorff L, Wagner CS (2008) International collaboration in science and the formation of a core group. J Informet 2(4):317–325
22. Liu HI, Huang MH (2022) Research contribution pattern analysis of multinational authorship papers. Scientometrics 127:1783–1800. https://doi.org/10.1007/s11192-022-04277-x
23. Luo Y (2022) Illusions of techno-nationalism. J Int Bus Stud:550–567. https://doi.org/10.1057/s41267-021-00468-5
24. Marginson S (2022) 'All things are in flux': China in global science. High Educ 83:881–910. https://doi.org/10.1007/s10734-021-00712-9
25. Marini G (2024) The megaproject race to fight Covid-19. A scientometric assessment of the National Institute of Allergy and Infectious Diseases (NIAID) International Funding. In: Cantoni F, Corazza L, De Nito E, Di Nauta P, Favari E (eds) Complexity and sustainability in megaprojects. MERIT 2023. Lecture Notes in Civil Engineering, vol 493. Springer, Cham. https://doi.org/10.1007/978-3-031-59703-9_16
26. Marini G, Mouritzen M (2024) The complexity of funding international research collaboration. An analysis of partnership and rivalry between the EU and China. J Stud Int Educ, 0(0). https://doi.org/10.1177/10283153241293379
27. Marini G, Yang L (2021) Globally bred Chinese talents returning home: An analysis of a reverse brain-drain flagship policy. Sci Public Policy 48(4):541–552
28. Miller C (2024) Global chip war for strategic semiconductors. Nat Rev Electr Eng 1:2–3. https://doi.org/10.1038/s44287-023-00005-6
29. Mody CC (2017) Academic centers and/as industrial consortia in American microelectronics research. Manag Organ Hist 12(3):285–303
30. Möllers N (2021) Making digital territory: Cybersecurity, techno-nationalism, and the moral boundaries of the state. Sci Technol Hum Values 46(1):112–138
31. Park S (2023) Semiconductors at the intersection of geoeconomics, technonationalism, and global value chains. Soc Sci 12(8):466
32. Peters MA (2022) Semiconductors, geopolitics and technological rivalry: the US CHIPS & Science Act, 2022. Educ Philos Theory 55(14):1642–1646
33. Quan W, Mongeon P, Sainte-Marie M, Zhao R, Larivière V (2019) On the development of China's leadership in international collaborations. Scientometrics 120(2):707–721. https://doi.org/10.1007/s11192-019-03111-1
34. Rüland AL (2023) Big science, big trouble? Understanding conflict in and around big science projects and networks. Minerva 61:553–580. https://doi.org/10.1007/s11024-023-09497-w

35. Shimizu H (2011) Scientific breakthroughs and networks in the case of semiconductor laser technology in the US and Japan, 1960s–2000s. Aust Econ Hist Rev 51(1):71–96
36. Shu F, Liu S, Larivière V (2022) China's research evaluation reform: what are the consequences for global science? Minerva 60:329–347. https://doi.org/10.1007/s11024-022-09468-7
37. Tang L (2024) Halt the ongoing decoupling and reboot US-China scientific collaboration. J Informet 18(2):101521
38. Tsay MY, Jou SJ, Ma SS (2000) A bibliometric study of semiconductor literature, 1978–1997. Scientometrics 49(3):491–509
39. Veletanlić E, Sá C (2020) Implementing the innovation agenda: a study of change at a Research Funding Agency. Minerva 58:261–283. https://doi.org/10.1007/s11024-020-09396-4
40. Wagner C (2008) The New Invisible College. Science for development. Brooking Institute Press
41. Wagner CS, Whetsell TA, Mukherjee S (2019) International research collaboration: novelty, conventionality, and atypicality in knowledge recombination. Res Policy 48(5):1260–1270. https://doi.org/10.1016/j.respol.2019.01.002
42. Wai S (2023) Inventor mobility and spillovers in the early semiconductor industry. Technol Soc 75:102331
43. Wang CC, Huang MH (2017) Knowledge spillovers among semiconductor companies with different technology positions and with different roles in the industry chain. In: ISSI, pp 741–752
44. Yoon SY, Jee SJ, Sohn SY (2021) Mapping and identifying technological coopetition: a multi-level approach. Scientometrics 126:5797–5817. https://doi.org/10.1007/s11192-021-04005-x

Using an Allocation Algorithm to Analyse Access to Cardiology Services in Italy

Domenico Marino[1], Giuseppe Quattrone[1(✉)], and Francesco Timpano[2]

[1] Mediterranea University of Reggio Calabria, Reggio Calabria, Italy
dmarino@unirc.it, gquattrone1@gmail.com
[2] Catholic University of Piacenza, Piacenza, Italy
francesco.timpano@unicatt.it

Abstract. This study proposes an allocation model to analyse the levels of access to cardiology services at a national level in Italy. Using an allocation algorithm implemented in R, the model considers demand from municipalities with a population over 65, available hospital facilities and transport distances. The results show how the model can provide detailed information on the distribution of demand in relation to distances, helping to improve accessibility to healthcare services. The analysis emphasises the importance of optimising patient flows and provides recommendations for the planning of healthcare resources.

1 Introduction

The efficiency of healthcare systems is a critical concern that has consistently attracted the attention of national and regional authorities. This interest stems not only from the direct impact on the population's health but also from the substantial costs required to ensure adequate care for all individuals. Designing a healthcare system that is both accessible and equitable is essential to achieving the principles of universality and equity outlined by the World Health Organization (WHO). The incorporation of these aspects into healthcare strategies allows for improved service quality through better resource planning while reducing emergencies that increase costs and patient health risks.

In recent years, many countries have initiated healthcare system reorganizations to manage rising costs, often resulting in resource reductions in various specialties, particularly in rural and less urbanized areas. These changes have amplified concerns about accessibility and equity, making it imperative to develop models that address these challenges effectively.

This study aims to address how the adoption of an allocation model incorporating a gravitational algorithm can enhance the planning and delivery of cardiology services in Italy by balancing demand, facility attractiveness, and territorial specificities to improve accessibility and resource optimization.

To explore this, this research focuses on developing a quantitative model to dynamically describe patient flows in the cardiology sector. This model considers both proximity to services and the utilization levels of individual facilities. By integrating spatial accessibility and allocation concepts, it offers a more comprehensive framework for understanding and addressing disparities in healthcare access.

E. Favari et al. (Eds.): MeRIT 2024, LNCE 600, pp. 244–252, 2025.
https://doi.org/10.1007/978-3-031-92797-3_16

The study is structured in multiple stages: a literature review on healthcare provision planning, particularly spatial accessibility and allocation models; a methodological explanation of patient movement simulation and attraction coefficient calculation; and an analysis of results derived from simulations. The findings provide insights that could guide healthcare resource planning and strategic decision-making for cardiology services in Italy.

2 Background

In the scientific literature, there are two main approaches to planning health services: spatial accessibility models and location-allocation models. The former focus on the principles of equity and service availability using geographical models to measure relative levels of access. The latter are developed to optimize service coverage via algorithms that simulate alternative hypotheses with respect to territorial health needs. Some authors have systematized the dimensions of service accessibility (Daskin and Dean 2004; Penchansky and Thomas 1981) by distinguishing between spatial and nonspatial factors influencing access to services. Among the most common methods used to measure accessibility are the Gaussian kernel density method (Guagliardo 2004; Okabe et al. 2009) and gravity models (Joseph and Bantock 1982; Schuurman et al. 2010). These models recognize the ability of spatial indicators to describe accessibility because of their ability to simultaneously consider both accessibility and availability. In particular, the distance between supply and demand is a key variable in these studies. Several authors have compared different ways of measuring this distance (Apparicio et al. 2008; Borruso and Borruso 2008; Shahid et al. 2009; Wong et al. 2012), indicating that access measured by road distance provides more accurate results than Euclidean or Manhattan distance does. From a computational point of view, a further complex element is the definition of the impedance factor used in gravitational models. Luo and Wang (Luo et al. 2003) tested how variations in primary care access outcomes are affected by the application of gravity models in urban versus rural areas. Recent studies note that policy decisions in the health sector, which are based on statistical analyses and indicators such as the ratio of providers to the population (Khan, 1992), always fail to adequately describe the real need for resources to be allocated on the ground (Fortney et al., 2000; Wang, 2012). The ability to respond to care depends not only on the activation of services in the territory but also on the ability of patients to reach these resources (Joseph, 1984; Khan and Bhardway, 1994, Marino and Quattrone, 2025). The complexity of mobility management increases further when one considers the territorial structure and the road network through which demand flows (Marino and Quattrone, 2019). Demands from rural areas with low population density and poor road infrastructure are rarely met with the same level of service as in urban areas (Pilkington et al., 2008). This phenomenon has prompted researchers to consider factors related to the isolation of territorial areas (Grzybowski et al., 2009; Kornelsen et al., 2010; Baray and Cliquet, 2013), where the specificities of the rural world may lead to the abandonment of hospital services in favour of less expensive home care. Such disparities in access to care can be reduced through better rationalization of the resources available in various areas, as well as through the introduction of new remote-controlled technologies for diagnosis and treatment (Marino et al., 2018, 2019, 2024). The analysis of the presented literature reveals that, despite the extensive body of research on healthcare planning models and service accessibility, many studies do not adequately address the role of the attractiveness of individual healthcare facilities in influencing patient flows. Moreover, the integration of dynamic algorithms that simultaneously account for demand, supply, and territorial specificities remains underexplored. This study aims to contribute to the debate by proposing a gravitational allocation algorithm capable of more accurately representing the dynamics of access to cardiology services and offering new perspectives for healthcare resource planning. The present work focuses on the measurement of spatial accessibility by considering the average road distance travelled to facilities with cardiology departments.

3 Materials and Methods

Data from various official sources were used to determine the size of demand in the model:

Population: In order to determine the size of the demand from the 7902 Italian municipalities, data from the ISTAT National Census were used to derive the population over 65 years of age. These data were extracted from the GEO-DEMO website.
Hospitalisation Facilities: The list of hospital facilities in Italy was acquired from the Ministry of Health website. For this simulation, an unlimited offered capacity is assumed.
Hospital admissions: The admissions were calculated on the basis of the case history of admissions for cardiological causes by interpolating the data on the resident population aged >65 in all municipalities of Italy.
Road Transport Network: Distances were calculated using origin-destination matrices processed by ISTAT.

The proposed model is based on a modified gravitational allocation algorithm to analyse the flow of patients to hospital facilities specialising in cardiology. This algorithm integrates several factors: the demand from municipalities with a population over 65 years, the availability of hospital facilities and the road distances required to reach these facilities. For the latter, the model uses origin-destination matrices to calculate the temporal distances between the municipalities and the healthcare facilities. Through this approach, the flow of patients is not based solely on geographical proximity but also takes into account the attractiveness of individual hospital facilities, thus obtaining a more realistic representation of access to healthcare services.

The formula adopted for the calculation of attraction coefficients is the one generally adopted for measuring levels of access to services through the use of gravity models (Joseph and Bantock 1982):

$$A_{ij} = \sum S_j / D_i d_{ij}^{\beta} \tag{1}$$

where $A_{i,j}$ represents the force exerted by the structure with load capacity S_j located at point j while D_i represents the demand for service originating from point *i*. From a spatial point of view, d_{ij} indicates the distance between the point of origin and the point of destination while the β coefficient measures the spatial friction between the population and the j-th service, assuming other factors, which influence the interaction, to be constant. By applying Vogel's algorithm (Reinfeld and Vogel 1958), the optimal allocation of care to individual facilities is provided using the criterion of minimising travel time. Using the attraction coefficients*A*, patients are allocated to the facilities that exert the greatest attraction force on them, providing for a non-penalisation mechanism designed to prevent patients from the most distant municipalities from being excluded a priori from the service.

For the formulation of the transport problem, the classical formulation was adopted, where the objective function was suitably modified by taking advantage of the coefficients of attraction that guide the displacement flows:

$$\max z = \sum_{i \in I} \sum_{j \in J} A_{ij} x_{ij} \tag{2}$$

$$\text{subject to}: \sum\nolimits_{j \in J} x_{ij} = D_i \quad \forall i = 1, .., m \tag{3}$$

$$\sum\nolimits_{i \in I} x_{ij} = S_j \, \forall j = 1, .., n \tag{4}$$

$$x_{ij} \geq 0 \, \forall ij \tag{5}$$

In the allocation phase, the objective function (2) determines the minimum total cost of a given transport. Constraints (3) and (4) require the number of admissions to be exactly equal to the number of patients the facilities are able to receive. This condition, balancing the system, is made possible by the creation of a fictitious destination, called "FR Facility" (Out-of-Region), which absorbs a demand equal to the surplus, if the total demand is greater than the total supply. Likewise, should total demand be lower than total supply, the resulting deficit would be absorbed by a fictitious out-of-region municipality, called 'FR Municipality'. Finally, constraint (5) requires that the flow of demand towards an inpatient facility is not negative. The implementation of the model in R allows a rapid simulation of patient flows and facilitates the analysis of the effects of variations in the availability of healthcare resources on the dynamics of access to cardiology services. The model also allows a predictive evaluation of the effects of changes in the structure of healthcare supply or in the organisation of the service itself. Furthermore, it enables decision-makers to test alternative scenarios concerning thedistribution of resources and thus improve strategic planning.

4 Results Obtained from Simulation

The main objective of the simulation conducted was to analyse the allocation of patient flows over the territory in relation to travel distances to cardiology departments, which represent the supply in our model. In this context, no restrictions were imposed beyond a certain distance. The propensity to move patients was determined exclusively by the beta decay coefficient of the gravitational formula used in the model. The results obtained offer a detailed insight into how different distance classes influence patient movement and care allocation.

Table 1. Results of Simulation.

Class of distance	Demand	Percentage demand (%)	Mean distance (min)	SD distance (min)
<15	119.016	1.8	10′	4′
15–30	615.195	9.4	23′	4′
30–60	2.362.164	36.1	47′	8′
60–120	2.156.415	32.9	91′	17′
>120	1.294.230	19.8	430′	219′
Tot	6.547.019	100.0		

The results of the simulation are summarised in the statistics table (Table 1), which presents the data for five distance classes.

The total aggregate demand was 6,547,019 units of allocated flows. The distance class 30–60 min recorded the highest flow with 2,362,164 units, representing 36.1% of the total allocation. This figure indicates that most patients tend to go to cardiology departments located at an intermediate distance, suggesting a preference for healthcare facilities that can be reached in a relatively short time. The distance classes 60–120 min and >120 min also contributed significantly to the allocation of flows, with 32.9% and 19.8% respectively. Interestingly, the presence of significant flows beyond 120 min can be explained by the lack of restrictions in our model regarding travel distances. In fact, the propensity to travel was influenced by the beta decay coefficient, which allowed patients to consider even more distant facilities if these offered specialised care or superior resources. The distance class <15 min, with a flow of 119,016 (or 1.8%), refers to patients residing in municipalities where a cardiology service is present and therefore represents the logistically optimal solution. Analysing the average distances associated with each class, it can be seen that the average for the <15-min class was 10 min, with a low standard deviation (4 min), suggesting a certain homogeneity in the assignments in this category. In contrast, the >120 min class showed a significantly higher mean distance (430 min) and high standard deviation (219 min), indicating greater variability in the routes travelled by patients. This reflects the ability of the model to capture patients' preferences for specialist care available at greater distances. The data analysis reveals some additional interesting considerations. Firstly, the availability of distant specialised services, combined with the propensity to travel without restrictions, highlights how the perceived quality of care is a determining factor. This could suggest a suboptimal distribution of cardiology services in closer areas, forcing patients to travel to more distant centres. This trend might indicate that facilities located in intermediate or nearby distances fail to meet demand, raising questions about their capacity or reputation. The concentration of demand in more distant ranges also reflects possible centralisation of services, where certain facilities act as main reference points, attracting a significant portion of patients even from long distances. While this phenomenon is understandable in terms of care quality, it may lead to overburdening of major centres. The issue of inequalities emerges strongly from the analysis of demand distribution. Patients living in rural or remote areas appear to be particularly disadvantaged, as they are forced to travel longer distances to access cardiology services. This territorial disparity reflects a potential disadvantage for those residing far from care centres, as travelling for extended periods can lead to delays in receiving care and, ultimately, worse clinical outcomes. It is clear that less-served areas could suffer from limited access, with significant consequences in terms of equity within the healthcare system.

5 Conclusions

This study represents an important first step in testing the proposed model for patient flow allocation based on travel distances to cardiology departments. The results obtained offer one of the first large-scale simulations analysing patient movements as a function of distance, providing valuable insights for healthcare resource planning and management. In light of the conclusions drawn from the previous data, several opportunities for improving the model emerge. One key area to address is the integration of capacity limitations for individual facilities. Currently, the model does not account for the fact that each cardiology department has a finite capacity in terms of patient intake. Including this constraint would lead to a more accurate reallocation of unmet demand flows, reflecting more realistically the real-world challenges of resource limitations and facility saturation. This could significantly alter the distribution of demand across regions, highlighting overburdened centres and helping balance the load more efficiently. Additionally, better calibration of attraction coefficients could further enhance the accuracy of the model. For instance, introducing a more precise variable defining the propensity for interregional travel, particularly over long distances, would allow the model to better capture patient movements, especially for those willing to travel for specialised care. This refinement could also address some of the inequalities observed, where patients in rural or underserved areas face greater barriers to accessing care, as indicated by the previous analysis. This study has laid the groundwork for future research in patient flow simulation, highlighting the complexity and dynamism of healthcare systems. The integration of additional variables, such as capacity constraints and more refined patient preferences, could transform the model into a crucial tool for optimising access to care and improving the efficiency of the healthcare system. These refinements would provide healthcare planners with better tools to manage resources, reduce inequalities, and ensure that patients receive timely and appropriate care across all regions.

References

1. Apparicio P, Abdelmajid M, Riva M, Shearmur R (2008) Comparing alternative approaches to measuring the geographical accessibility of urban health services: distance types and aggregation-error issues. Int J Health Geogr 7(1):7
2. Baray J, Cliquet G (2013) Optimizing locations through a maximum covering/p-median hierarchical model: maternity hospitals in France. J Bus Res 66(1):127–132
3. Borruso G (2008) Network density estimation: a GIS approach for analysing point patterns in a network space. Trans GIS 12(3):377–402

4. Crooks VA, Schuurman N (2012) Interpreting the results of a modified gravity model: examining access to primary health care physicians in five Canadian provinces and territories. BMC Health Serv Res 12(1):230
5. Cuomo C, Marino D (2019) The LOCAL WORK PLANS (LWP) and Territorial Economic System (TES): Assessment and Evaluation. In: Gil-Lafuente, J., Marino, D., Morabito, F. (eds) Economy, Business and Uncertainty: New Ideas for a Euro-Mediterranean Industrial Policy. AEDEM 2017. Studies in Systems, Decision and Control, vol 180. Springer, Cham. https://doi.org/10.1007/978-3-030-00677-8_19
6. Cueto M (2004) The origins of primary health care and selective primary health care. Am J Public Health 94(11):1864–1874
7. Daskin MS, Dean LK (2004) Location of health care facilities. In: Sainfort F, Brandeau M, Pierskalla W (eds) The handbook of OR/MS in health care: a handbook of methods and applications. Kluwer Academic, Boston, pp 43–76
8. Field K (2000) Measuring the need for primary health care: an index of relative disadvantage. Appl Geogr 20(4):305–332
9. Fyer GE, Drisko J, Krugman RD, Vojir CP, Prochazka A, Miyoshi TJ, Miller ME (1999) Multi-method assessment of access to primary medical care in rural Colorado. J Rural Health 15(1):113–121
10. Galvao RD, Acosta Espejo LG, Boffey B (2002) A hierarchical model for the location of perinatal facilities in the municipality of Rio de Janeiro. Eur J Oper Res 138(3):495–517
11. Grzybowski S, Kornelsen J, Schuurman N (2009) Planning the optimal level of local maternity service for small rural communities: a systems study in British Columbia. Health Policy 92(2):149–157
12. Grzybowski S, Stoll K, Kornelsen J (2011) Distance matters: a population based study examining access to maternity services for rural women. BMC Health Serv Res 11(1):147
13. Guagliardo MF (2004) Spatial accessibility of primary care: concepts, methods and challenges. Int J Health Geogr 3(1):3
14. Hillier FS (2015) Introduction to operations research, 10th edn. McGraw-Hill, New York, NY
15. Joseph AE (1984) Accessibility and utilization: geographical perspectives on health care delivery. Harper & Row, New York
16. Joseph AE, Bantock PR (1982) Measuring potential physical accessibility to general practitioners in rural areas: a method and case study. Soc Sci Med 16(1):85–90
17. Khan AA (1992) An integrated approach to measuring potential spatial access to health care services. Socioecon Plann Sci 26(4):275–287
18. Khan AA, Bhardwaj SM (1994) Access to health care: a conceptual framework and its relevance to health care planning. Eval Health Prof 17(1):60–76
19. Kornelsen J, Kotaska A, Waterfall P, Willie L, Wilson D (2010) The geography of belonging: the experience of birthing at home for First Nations women. Health Place 16(4):638–645
20. Levaggi R, Menoncin F (2008) Fiscal federalism, patient mobility and soft budget constraint in Italy. Polit Econ 3:367
21. Liu L, Hader J, Brossart B, White R, Lewis S (2001) Impact of rural hospital closures in Saskatchewan, Canada. Soc Sci Med 52(12):1793–1804
22. Luo W (2004) Using a GIS-based floating catchment method to assess areas with shortage of physicians. Health Place 10(1):1–11
23. Luo W, Wang F (2003) Measures of spatial accessibility to health care in a GIS environment: synthesis and a case study in the Chicago region. Environ Plann B Plann Des 30(6):865–884
24. Marianov V, Serra D (2011) Location of multiple-server common service centers or facilities, for minimizing general congestion and travel cost functions. Int Reg Sci Rev 34(3):323–338

25. Marino D, Quattrone G (2019) A proposal for a new index to evaluate hospital resource allocation: the case of Italian NHS rationalisation. Eur Res Manag Bus Econ 25(1):23–29. https://doi.org/10.1016/j.iedeen.2018.10.001
26. Marino D, Cananzi D, Aragona F (2024) Artificial intelligence and ethics challenges of technologies. In: Ethics and artificial intelligence. Studies in systems, decision and control, vol 519. Springer, Cham. https://doi.org/10.1007/978-3-031-50902-5_3
27. Marino D, Quattrone G (2025) Modelling health mobility for equity distribution services in the emergency health sector. Ann Reg Sci 74, 46. https://doi.org/10.1007/s00168-025-01373-7
28. Marino D, Miceli A, Quattrone G (2018) Artificial intelligence and changing paradigm in healthcare. In: Lafuente JG, Marino D, Morabito FC (eds) Economy, business and uncertainty. Palgrave Macmillan Springer Nature, Cham
29. Domenico M, Miceli A, Carlizzi DN, Quattrone G, Sancin C, Turchi M (2019) Telemedicine and Impact of Changing Paradigm in Healthcare. In: Calabrò Francesco Della Spina Lucia Bevilacqua Carmelina, New Metropolitan Perspectives. Springer Nature, Berlin
30. Marino D et al (2023) Artificial Intelligence as a disruption technology to build the harmonic health industry. Procedia Comput Sci 217:1354–1359. https://doi.org/10.1016/j.procs.2022.12.333
31. Masters SH, Burstein R, Amofah G, Abaogye P, Kumar S, Hanlon M (2013) Travel time to maternity care and its effect on utilization in rural Ghana: a multilevel analysis. Soc Sci Med 93:147–154
32. Matsumoto M, Ogawa T, Kashima S, Takeuchi K (2012) The impact of rural hospital closures on equity of commuting time for haemodialysis patients: simulation analysis using the capacity-distance model. Int J Health Geogr 11(1):28
33. McArthur DP, Gregersen FA, Hagen TP (2014) Modelling the cost of providing ambulance services. J Transp Geogr 34:175–184
34. McGrail MR, Humphreys JS (2009) Measuring spatial accessibility to primary care in rural areas: improving the effectiveness of the two-step floating catchment area method. Appl Geogr 29(1):533–541
35. Okabe A, Satoh T, Sugihara K (2009) A kernel density estimation method for networks, its computational method and a GIS-based tool. Int J Geogr Inf Sci 23(1):7–32
36. Penchansky R, Thomas JW (1981) The concept of access: definition and relationship to consumer satisfaction. Med Care 19(2):127–140
37. Pilkington H, Blondel B, Carayol M, Breart G, Zeitlin J (2008) Impact of maternity unit closures on access to obstetrical care: the French experience between 1998 and 2003. Soc Sci Med 67(10):1521–1529
38. Pilkington H, Blondel B, Papiernik E, Cuttini M, Charreire H, Maier RF, Petrou S, Combier E, Künzel W, Bréart G et al (2010) Distribution of maternity units and spatial access to specialised care for women delivering before 32 weeks of gestation in Europe. Health Place 16(3):531–538
39. Reinfeld NV, Vogel WR (1958) Mathematical programming. Prentice-Hall, Englewood Cliffs, NJ
40. Revelle C, Marks D, Liebman JC (1970) An analysis of private and public sector location models. Manag Sci 16(11):692–707
41. Schuurman N, BéRubé M, Crooks VA (2010) Measuring potential spatial access to primary health care physicians using a modified gravity model. Can Geogr 54(1):29–45
42. Shahid R, Bertazzon S, Knudtson ML, Ghali WA (2009) Comparison of distance measures in spatial analytical modeling for health service planning. BMC Health Serv Res 9(1):200
43. Wang F (2012) Measurement, optimization, and impact of health care accessibility: a methodological review. Ann Assoc Am Geogr 102(5):1104–1112

44. Wong LY, Heng BH, Cheah JTS, Tan CB (2012) Using spatial accessibility to identify polyclinic service gaps and volume of under-served population in Singapore using Geographic Information System. Int J Health Plann Manage 27(3):e173–e185

Author Index

E. Favari et al. (Eds.): MeRIT 2024, LNCE 600, p. 253, 2025.
https://doi.org/10.1007/978-3-031-92797-3

The manufacturer's authorised representative in the EU is Springer Nature Customer Service Centre GmbH, Europaplatz 3, 69115 Heidelberg, Germany. If you have any concerns regarding our products, please contact ProductSafety@springernature.com

Printed and bound by CPI Group (UK) Ltd, Croydon, CR0 4YY
05/10/2026
02228126-0004